EXERCISES
IN
ALGEBRA
AND
TRIGONOMETRY

D.R. OLSON

SECOND EDITION

Preface

This book contains exercises that a student can complete to learn algebra and trigonometry. I designed it as a companion to another book of mine, namely *Algebra and Trigonometry*, but the exercises herein are typical for the subjects and do not pertain exclusively to that particular text.

D.R. Olson

17 June 2023

Contents

Part One

Part Two

PART ONE

Exercises in Algebra and Trigonometry

Set 1: Algebraic Expressions

1. Evaluate each expression for $a = 3$, $b = -2$, and $c = 7$.

A. $4a$

F. $12/(3c - 6a)$

B. $b - a$

G. $-b(20 - 8a)$

C. $2a + c$

H. $2ab + 3bc + 4ac$

D. $-c - 3b - 11$

I. $(c - b)(c + b)$

E. $7(4a + 5b)$

J. $31 - 2[b(5 + 9a)]$

2. Determine if the number 3 is a zero of the given expression.

A. $-6 + 2x$

C. $b^3 - 5b^2 + 4b + 7$

B. $m + 5$

D. $(n^2 + 2n - 15)/91$

Set 2: Sets

1. Define a set that contains the elements described. Choose an appropriate capital letter to represent the set.

   A. the colors on the national flag of Canada

   B. the countries on the Baltic Sea

   C. the last names of the most recent five Presidents of the United States

   D. the ingredients in chocolate chip cookies

   E. the even integers between 11 and 25

   F. the symbols of the noble gases

   G. the most recent seven years in which Halley's Comet was visible from Earth

   H. the five longest rivers in North America

2. Determine whether the given statement is True or False.

   A. $7 \in \{1, 3, 7, 8\}$

   B. $\{2, 3\} \subset \{1, 3, 7, 8\}$

   C. $4 \in \{1, 3, 7, 8\}$

   D. $\{1, 7, 8\} \in \{1, 3, 7, 8\}$

   E. $8 \subset \{1, 3, 7, 8\}$

   F. $\{3\} \subset \{1, 3, 7, 8\}$

3. Suppose $L = \{a, b, c\}$, $M = \{a, b, c, d, e\}$, and $N = \{b, d, f, g\}$. List the elements in each set.

   A. $L \cup N$

   B. $M \cap N$

   C. $\{(x, y) \mid x \in N, y \in L\}$

   D. $M \backslash L$

   E. $\{n \in N \mid n$ is one of the first four letters of the alphabet $\}$

   F. $M \cap N \cap L$

   G. $L \backslash M$

   H. $\{x \in L \mid x \in N\}$

   I. $L \cup N \cup M$

Set 3: Real Numbers

1. Of the sets **N**, **Z**, **Q**, and **R**, determine the smallest set that contains the given number.

A. 8.625

G. 2,000,000

B. π

H. $\sqrt{2}$

C. $-11/3$

I. 9.1

D. -6

J. $20/5$

E. $\sqrt{9}$

K. 0π

F. 0

L. $5 + e$

2. Provide expressions that represent the integers described.

A. the higher of two consecutive integers where m represents the lower of the two

B. the lower of two consecutive odd integers where n represents the higher of the two

C. the highest of three consecutive even integers where c represents the lowest of the three

D. the lowest of eight consecutive integers where $5d - 11$ represents the highest of the eight

Set 4: Interval Notation

1. Write each set in interval notation.

A. $\{\, x \in \mathbf{R} \mid -7 < x \le -4 \,\}$ E. $\{\, x \in \mathbf{R} \mid 5 \le x \,\}$

B. $\{\, x \in \mathbf{R} \mid x > -1 \,\}$ F. $\{\, m \in \mathbf{R} \mid -12 < m < -10 \,\}$

C. $\{\, y \in \mathbf{R} \mid 100 \le y \le 200 \,\}$ G. $\{\, w \in \mathbf{R} \mid w \le -2 \,\}$

D. $\{\, x \in \mathbf{R} \mid 9 > x \,\}$ H. $\{\, x \in \mathbf{R} \mid 8 > x \ge 3 \,\}$

2. Write each set in set builder notation.

A. $[2, 7]$ E. $[9, \infty)$

B. $(-\infty, 5)$ F. $(-10, 0)$

C. $(-1, 100]$ G. $[6, 17)$

D. $(-8, \infty)$ H. $(-\infty, -3]$

3. Simplify each expression.

A. $(5, 11] \cap (7, 14)$ E. $(1, 10) \cup [4, 9]$

B. $(-11, -5] \cup [-8, -3]$ F. $(-\infty, -4) \cap [-2, 5]$

C. $[-7, \infty) \cap (2, 6)$ G. $(3, 6] \cap [6, 11)$

D. $(-\infty, 15] \cap [10, 20)$ H. $(-\infty, 4] \cup (3, \infty)$

Set 5: The Real Number Line

1. Draw the graph of each set of real numbers.

A. $\{-8, -2, 0, 5\}$

H. $[9, \infty)$

B. $[2, 7]$

I. $\{-5, -3, -1, 1, 3, 5\}$

C. $(-4, \infty)$

J. $[8, 17)$

D. $(-\infty, -5]$

K. $(-10, 0)$

E. $(-1, 1]$

L. $(-\infty, 100)$

F. $\{\, x \in \mathbf{Z} \mid 1 < x \leq 10 \text{ AND } x \text{ is even} \,\}$

G. $\{\, n \in \mathbf{Z} \mid 5 \leq n < 20 \text{ AND } n \text{ is a multiple of } 3 \,\}$

Set 6: Algebraic Statements

1. Determine whether the given statement is true or false, or if its veracity depends on the value of the variable it contains.

A. $7 - 5 = -2$

D. $4[1 + (2 - 5)/6] = 2$

B. $7x = 21$

E. $5/(x - 3) = 0$

C. $m^2 < 3m - 5$

F. $2c^2 + 5c + 3 < 0$

2. Determine whether the given numbers are solutions of the given equation.

A. $8 = 2x$; $4, -3$

C. $x^2 - 3x = 4$; $-2, -1, 4$

B. $3x - 5 > 1$; $0, 2, -3$

D. $y^2 < 5$; $-2, 1, 3, 7$

Set 7: Properties of Operations on Real Numbers

1. Identify the additive and multiplicative inverses of each real number.

A. 8

D. -1

B. $2/7$

E. 0

C. e

F. $-4/3$

2. Identify the property that justifies each statement.

A. $5 \cdot \frac{1}{5} = 1$

G. $(b \cdot 2) + (c \cdot 5) = (c \cdot 5) + (b \cdot 2)$

B. $5 + 8 = 8 + 5$

H. $(-2 \cdot 9) \cdot 3 = -2 \cdot (9 \cdot 3)$

C. $-7 + 0 = -7$

I. $-3 + 3 = 0$

D. $11 \cdot (9 + x) = 11 \cdot 9 + 11 \cdot x$

J. $1 \cdot 4 = 4$

E. $m(6) = 6(m)$

K. $v + (13 + 2) = (v + 13) + 2$

F. $-9 + (7 + 0) = -9 + 7$

L. $5 \cdot (2 + 11) = 5 \cdot (11 + 2)$

3. Simplify each numerical expression using properties of addition and multiplication. Justify each step and do not use a calculator.

A. $587(-25) + 587(24)$

C. $(229 \cdot 500) \cdot 2$

B. $[8{,}721 + (-64{,}352)] + 64{,}352$

D. $37(96) + 37(4)$

4. Use an example to show that subtraction is not associative.

5. Use an example to show that division is not commutative.

6. Determine whether the given statement is True or False.

A. "Identity elements and inverses exist under subtraction and division."

B. "Every real number has a multiplicative inverse."

C. "Every real number has an additive inverse."

D. "An operation can have more than one identity element."

Set 8: Powers and Exponents

1. Evaluate each numerical expression.

A. 4^3

F. $(0.6)^2$

B. $(50/25)^4$

G. 8^0

C. $(-5)^2$

H. -9^2

D. -2^0

I. $3^3 - 2^4$

E. $(1/5)^3$

J. $-7^2 + (-7)^2$

2. Rewrite each expression as an expanded product of factors without exponents, then simplify the product into a compact form with exponents.

A. $3x^2 \cdot 5x$

C. $(12b^7)/(3b^2)$

B. $(7m^3)(2m^2)(8m^6)$

D. $(6x)(10x^{11})/(4x^5)$

3. Evaluate each expression using the given values of the variables.

A. $7x^3$ for $x = -2$

B. $a^2 - b^4$ for $a = -8$ and $b = -1$

C. $4x^5 + 10y^2 - 2z^3$ for $x = 2$, $y = -5$, and $z = -3$

D. $-8m^3 + 12n^2$ for $m = 1/2$ and $n = 2/3$

4. Evaluate the expression $2^9 \cdot 5^6$.

Set 9: Like Terms

1. Simplify each expression.

A. $5x - 4x - 7x$

B. $10mn^2 + 3mn^2$

C. $\frac{3}{4}z - \frac{1}{5}z + \frac{1}{2}z + \frac{7}{3}z$

D. $-x^3y^5 - 8x^3y^5 + 20x^3y^5$

E. $2b + 5b + 10c - 3c$

F. $3v - 5w - 7v + 2v + 8w$

Set 10: Simplifying Algebraic Expressions

1. Simplify each expression.

A. $(8x^3)(3x)$

E. $(4y)(-5y^2)(2y^7)$

B. $2m + 11m - 6m - 3m$

F. $-7p + 2q + 9q - 3p$

C. $3(x + 9) - 5(x - 2)$

G. $5x(x - 2) - 10(x + 7) + 3$

D. $y(y^2 + 4y - 3) + 2(8 - 5y)$

H. $-7(2x - 5y + 3z) + 4(5x - 9y - z)$

I. $6x^3 + 24x^2 - 2x^2 - 8x + 9x^2 + 36x - 3x - 12$

J. $m(n + 6) + 3mn - 10 - 2n(8 - 4m - n) + 4n^2 + 7n$

2. Simplify each expression and then evaluate the result for the given values of the variables.

A. $4(x + 3) - 5(x + 6)$ for $x = 2$

B. $-3(2x - y) + 2(3x + 5y)$ for $x = 15$ and $y = -8$

C. $8(3 - x) - 5(x + 2) + 12(x - 1) + x - 2$ for $x = 7$

D. $7(x - 4) - x(1 + x) - 10x^2 + 12 + 6(x^2 - x)$ for $x = -3$

E. $x(y - 2z + 7) - 2y(8x - 3z + 2) - 3z(7 - 2x)$ for $x = 1$, $y = 3$, $z = -1$

F. $x(x - y) + y(x - y)$ for $x = 5$ and $y = 4$

Set 11: Properties of Equality and Inequality

1. Identify the property that justifies each statement.

    A. $x - 5 = 2$ if and only if $(x - 5) + 5 = 2 + 5$

    B. $-m > 8$ if and only if $(-1)(-m) < (-1)(8)$

    C. if $n = 2m$ and $10m - 7 = 3n$, then $10m - 7 = 3(2m)$

    D. $2 = 2$

    E. if $x < y$ and $y < -9$, then $x < -9$

    F. $4y = 12$ if and only if $\frac{1}{4}(4y) = \frac{1}{4}(12)$

    G. if $b = a$ and $a = 11$, then $b = 11$

    H. $c + 11 < 20$ if and only if $(c + 11) - 11 < 20 - 11$

    I. if $x = 7$, then $7 = x$

    J. $5z < -20$ if and only if $\frac{1}{5}(5z) < \frac{1}{5}(-20)$

Set 12: Linear Equations in One Variable

1. Solve each equation.

A. $x - 3 = 8$

L. $7 = 2 + x$

B. $5y = 30$

M. $\frac{1}{4}x = 9$

C. $-\frac{5}{8}n = \frac{25}{4}$

N. $2q = \frac{1}{3}$

D. $-3x + 2 = 14$

O. $x/5 = 11$

E. $4m/3 = -1/7$

P. $10 - 2x = 4 + x + 6 - 3x$

F. $7 - 3x = 5x + 31$

Q. $9x - 16 = 7x$

G. $-11c - 4 = -5c - 9 - c$

R. $3.4m - 6.7 = 1.1m + 2.5$

H. $3x - 8 = 14 + 3x$

S. $10x + 3 = 3$

I. $2/3 - 6m = -3/4 - m$

T. $-y + 8 = 2$

J. $8x - 5 = 6 + 8x - 11$

U. $x + 4 = -1 + x$

K. $4.9x + 6.2 = 40.5$

V. $6x - 5 - x - 2 = 17 - 2x - 4 - 3x$

2. Evaluate each expression using the given condition.

A. $3x + 6y$ if $x + 2y = 7$

C. $4a - b$ if $8a - 2b = 18$

B. $5n - 3m$ if $3m - 5n = 11$

D. $7y + 4x$ if $4x + 7y = -3$

Set 13: From English to Algebra

1. Translate each phrase into an algebraic expression.

A. the difference of 8 and x

J. 12 decreased by x

B. the quotient of m and 6

K. five times n

C. twice x

L. x minus 17

D. the sum of y and 11

M. m more than n

E. p divided by negative three

N. a subtracted from b

F. n less than 7

O. 2 divided into w

G. x plus y

P. x multiplied by 20

H. the product of 14 and h

Q. 11 added to x

I. k increased by 6

R. y less 1

2. Answer each question with an algebraic expression.

A. The population of Seattle is s. What would be twice the population?

B. Bill weighs b pounds. Carl weighs 15 pounds more than Bill. How much does Carl weigh?

C. Earl can run x meters in 60 seconds. How fast can Earl run?

D. Suppose x is an odd integer. What is the next higher odd integer?

E. Fred sells donuts for d cents apiece. How many cents does he earn from selling 10 donuts?

F. The high temperature on Sunday was t degrees. On Monday, the high temperature dropped eleven degrees from the day before. What was the high temperature on Monday?

G. Sixteen people share equally a pizza of mass x grams. How much pizza does each person get?

H. Mike had m pesos before a thief stole 200 pesos from him. How much money does Mike have now?

Set 14: Word Problems: Linear Applications

1. Solve each word problem.

    A. The product of 7 and a certain number is 28. Determine the number.

    B. The difference of a certain number and 3 is 15. Determine the number.

    C. $2.50 plus twice the price of a hamburger is $13. What is the price of the hamburger?

    D. Five times the length of a house minus seven yards is 88 yards. What is the length of the house?

    E. The quotient of the number of pages in a book and 8 is 53. How many pages does the book have?

    F. Seven years ago, Bill was two years old. How old is Bill now?

    G. Nine less than five times a number is seventy-six. Find the number.

    H. The number of people who voted in a recent election in California exceeded the number of registered voters in the region by 95,830. The number of registered voters in the region was 44,271. How many people voted in the election?

    I. Find two consecutive odd integers whose sum is 88.

    J. The perimeter of a square is 72 feet. Determine the length of the sides of the square.

    K. Hiro has a jar that contains only dimes and nickels, where the number of dimes is three times the number of nickels. The total amount of money in the jar is $6.65. How many nickels does the jar contain?

    L. Four people divide equally the beans in a bag. The sum of 24 and the number of beans that each person receives is 81. How many beans were in the bag?

    M. A piece of string is cut into two pieces. One piece is six inches longer than the other piece. The original length of the string was 110 inches. What are the lengths of the pieces of string?

    N. A bicycle and sidecar cost $230 in total. The bicycle cost $22 less than twice the cost of the sidecar. Find the price of the sidecar.

Set 15: Linear Equations with Parentheses

1. Solve each equation.

A. $3(x + 6) = 27$

E. $4(w - 2) = 3w + 8$

B. $-5(x + 9) = -35$

F. $6(3 - m) - 5 = 2m - 3$

C. $9(v + 2) - 5(v - 1) + 2 = 15$

G. $5(x - 2) + 2(x + 9) = 8$

D. $-3(x + 7) = 10(x - 3) - 4$

H. $-(n - 8) - 3n - 5 = 6 - 10n$

I. $2[7(x + 11) - 5(12 - x)] - 5x = 13 - 3(-4x + 1) + 10$

2. Solve each word problem.

A. The sum of an odd number and three times the next largest odd number is 50. Find the numbers.

B. The sum of two numbers is 17. The sum of four times the smaller number and three times the larger number is 58. Find the numbers.

C. Tony ate 75 single-topping pizzas last year. Each topping was either pepperoni, sausage, or beans. The number of pizzas he ate with sausage was twice the number with pepperoni, and the number with beans was seven more than the number with pepperoni. Find the number of pizzas he ate with pepperoni.

D. Lana sold 38 tickets to her piano recital. She sold some tickets to members of her tribe for $3 and some tickets to her friends for $8. How many tickets did she sell to her friends if she earned a total of $149.

E. Find two consecutive odd integers where the difference of five times the smaller number and two times the larger number is 47.

F. Find three consecutive even integers where the three times the sum of the numbers is 252.

Set 16: Linear Equations with Fractions

1. Solve each equation.

A. $3k/5 = 4$

B. $6 - 2x/7 = -3$

C. $(n+5)/2 = 11/3$

D. $\frac{3}{4}x + \frac{1}{2} = \frac{7}{5}$

E. $\dfrac{c-5}{7} = \dfrac{1-3c}{2}$

F. $\dfrac{4x+5}{4} + \dfrac{3-9x}{8} = -9$

G. $3m/8 = 2/3$

H. $4w/9 + 2 = 5/11$

I. $3v - \frac{5}{6} = 9 + \frac{7}{4}v$

J. $2t + \frac{1}{6}t - \frac{5}{4} = 8 + \frac{10}{3}t$

K. $\dfrac{9x}{7} = \dfrac{4x+5}{3} - \dfrac{x-2}{21}$

L. $\dfrac{5-x}{4} - \dfrac{3x}{8} = -\dfrac{1-x}{2} - \dfrac{9x-14}{8}$

2. Solve each word problem.

A. The difference of two-fifths of a number and one-fifth of the same number is nine-tenths. Find the number.

B. The quotient of the difference of five and a number and two equals three-sevenths. Find the number.

C. The sum of 12° and three-fourths of the measure of the complement of an angle is 75°. Find the measure of the angle.

D. The sum of one-third of the measure of the supplement of an angle and the measure of the angle itself is 142°. Find the measure of the angle.

Set 17: Subtraction and Division of Terms

1. Solve each equation.

A. $x - 8 = 14$

F. $6 - w = 7$

B. $2t + 5 = -13$

G. $3(2m + 1) = 5(m - 7)$

C. $7x = 28$

H. $k/9 = 11$

D. $(x - 4)/6 = -5x/3$

I. $7(1 - 2x) = 4x + 9$

E. $4x = (10 - x)/5$

J. $5 + v/6 = -8 - v$

2. Use the formula $F = \frac{9}{5}C + 32$ in the exercises below. This formula relates degrees Fahrenheit (F) to degrees Celsius (C).

A. Find F where $C = 100$.

C. Find F where $C = 15$

B. Find C where $F = 32$.

D. Find C where $F = 86$.

E. Determine the temperature at which the number of degrees Fahrenheit equals the number of degrees Celsius.

3. Rewrite each repeating decimal in the form a/b where $a \in \mathbf{Z}$ and $b \in \mathbf{N}$.

A. $0.\overline{5}$

E. $0.0\overline{2}$

B. $0.\overline{417}$

F. $8.8\overline{3}$

C. $6.9\overline{29}$

G. $0.7\overline{186}$

D. $3.45\overline{6}$

H. $1.93\overline{41}$

Set 18: Linear Inequalities in One Variable

1. Solve each inequality.

A. $x - 5 > 9$

H. $w + 6 \geq 2$

B. $3t - 5 \leq -11$

I. $2(4m - 1) < 3(m + 6)$

C. $9x \geq 27$

J. $-5x > 15$

D. $k/4 < 11$

K. $(7 - x)/2 \leq -3x/5$

E. $-w/10 > 2$

L. $11 + 2x \geq 9$

F. $7m + 10 < 8(m - 3) + 15$

M. $3 - v/6 \leq -2 - v$

G. $1 < 3x + 7 \leq 22$

N. $-8 \leq 2 - 5x < 27$

2. Solve each word problem.

A. Eleven less than seven times a number is greater than seventeen. State the possible values of the number in interval notation.

B. Boris took four exams and earned scores of 78, 91, 84, and 74. What is the lowest score that Boris can attain on the fifth exam in order for the average of all five scores to be at least 85?

C. James has $50 to spend in total on three books. He selects one book that costs $18 and another book that costs $13. What is the maximum price of the third book that he could select?

D. A baseball team has won 30 games and lost 24 games. The team will play 36 more games this season. Of these remaining games, what is the minimum number that the team must lose in order to ensure that it has won fewer than half of all the games it plays this season?

Set 19: Compound Inequalities

1. State the set of solutions of each compound inequality.

A. $x \geq -5$ AND $x < 3$

E. $x < -7$ OR $x \leq -3$

B. $x < 1$ OR $x > 4$

F. $x \leq -2$ AND $x < 10$

C. $x \geq -11$ OR $x \leq 2$

G. $x > 8$ OR $x > 12$

D. $x < 5$ AND $x \geq 6$

H. $x \geq 9$ AND $x \geq 7$

2. Solve each compound inequality.

A. $3x < 6$ AND $x + 4 \leq 12$

E. $9 - x < 2$ OR $-5x < 5$

B. $5 + x < 8$ AND $x/2 \geq 5$

F. $4x + 1 > x - 17$ OR $-8x > 8$

C. $2x + 3 > -7$ AND $-x/4 \geq -3$

G. $x \leq -6 - x$ OR $7x < -35$

D. $x/4 - 7 \geq -5$ OR $-6x > -18$

H. $5(x - 5) > 3(x - 7)$ AND $x + 1 > 4$

3. Solve each word problem.

A. David took three exams and earned scores of 93, 81, and 87. He would like to finish the course with a grade of B, which he will earn if his average score on all four of the exams in the course is greater than or equal to 80 and less than 90. In what interval must his score on the fourth exam lie in order for him to earn a grade of B?

B. Eric took the same three exams that David did and earned scores of 83, 72, and 68. Eric abhors mediocrity and would be happy to finish the course with any grade other than C, which he will earn if his average score on all exams is greater than or equal to 70 and less than 80. In what interval(s) must his score on the fourth exam lie in order for him to avoid the C? Assume scores can range from 0 to 100, inclusive.

Set 20: Absolute Value

1. Evaluate each expression.

    A. $|-10|$ C. $|7|$

    B. $|0|$ D. $|\pi - 5| - 5$

2. Solve each equation.

    A. $|x| = 8$ I. $|w| = -5$

    B. $|m| = 0$ J. $|x| = 3$

    C. $|5x| = 15$ K. $|4x| = 24$

    D. $|x| + 1 = 17$ L. $|w| - 5 = -3$

    E. $|x/3| = 10$ M. $|-y/8| = 2$

    F. $|4x - 3| = 5$ N. $|5 - 3x| = 44$

    G. $\left|\dfrac{x + 6}{5}\right| = 3$ O. $\left|\dfrac{4n - 1}{7}\right| = 5$

    H. $|2x - 6| + 13 = 21$ P. $|1 - 5x| - 2 = 12$

3. Solve each inequality.

    A. $|x| < -3$ I. $|x| \geq -4$

    B. $|k| \leq 5$ J. $|w| > 0$

    C. $|7x| > 28$ K. $|6x| \leq 42$

    D. $|x| - 7 \geq -1$ L. $|x| + 2 < 23$

    E. $|-v/7| \leq 3$ M. $|9 - 2x| \geq 7$

    F. $|9x + 1| > 8$ N. $|x/9| < 4$

    G. $\left|\dfrac{4n + 1}{5}\right| \geq 5$ O. $\left|\dfrac{11 - x}{2}\right| \leq 8$

    H. $|2 - 6x| - 3 < 13$ P. $|4x - 7| + 10 > 23$

Set 21: Distance and Midpoint Formula, One Dimension

1. Given the coordinates of points P and Q on the number line, find:
   - the distance $d(P,Q)$ between the points and
   - the coordinate of the midpoint of the line segment PQ.

A. $P : 5$, $Q : 9$

E. $P : -6$, $Q : -5$

B. $P : -2$, $Q : 11$

F. $P : 15$, $Q : 7$

C. $P : -1$, $Q : -7$

G. $P : -13$, $Q : 0$

D. $P : 0$, $Q : 10$

H. $P : 3$, $Q : -3$

Set 22: Geometric Formulas

1. Determine the specified quantities.

    A. the area and perimeter of a rectangle with length 5 and width 8

    B. the area and circumference of a circle with radius 7 cm

    C. the volume and surface area of a sphere with radius 4 ft

    D. the volume and surface area of a cylinder with height 6 in and radius 2 in

2. Solve each word problem.

    A. The sides of a certain stop sign are 12.4 inches long. What is the perimeter of the sign? [Hint: A stop sign has the shape of an octagon.]

    B. Alison wants to fill a box with pudding and give it to her friend as a present for Christmas. The box measures 9 cm by 10 cm by 4.5 cm. How much pudding can she fit into the box?

    C. Bessie finds a large conical icicle. Its height is 18 inches and its radius is 5 inches. What is the volume of the icicle?

    D. Craig builds a fence around a rectangular lot with dimensions 50 yards by 70 yards. What is the total length of the fencing?

    E. The floor of a house has the shape of a parallelogram with base 30 ft and height 50 ft. How many square feet of carpet are required to cover the floor?

    F. Fred forms a ring on a piece of paper by drawing two concentric circles and then shading the region between the circles. The larger circle has radius 5 cm and the smaller circle has radius 4 cm. What is the area of the ring?

    G. Igor absentmindedly drives around a traffic circle six times. The diameter of the circle is 3 kilometers. How far did Igor drive?

    H. The capstone of a certain pyramid also takes the shape of a pyramid, but is made of solid gold. The height of the capstone is 8 inches and its base has area 144 square inches. How much gold is in the capstone?

    I. What is the surface area of a basketball with radius 4.75 inches?

J. Tom cuts a cardboard tube from one end to the other and then unrolls it to form a rectangle. If the tube had length 19 cm and radius 3 cm, then what is the area of the rectangle?

3. Use the given formula and values to solve for the specified variable.

A. Solve $d = rt$ for r, where $d = 120$ and $t = 3$.

B. Solve $F = ma$ for a, where $F = 1200$ and $m = 50$.

C. Solve $i = Prt$ for P, where $i = 28$, $r = 0.07$, and $t = 4$.

D. Solve $A = P + Prt$ for t, where $A = 3{,}080$, $P = 2{,}000$, and $r = 0.03$.

E. Solve $z = \dfrac{x - \mu}{\sigma}$ for x, where $z = -1.5$, $\mu = 75$, and $\sigma = 8$.

F. Solve $A = (x_1 + x_2 + x_3 + x_4 + x_5)/5$ for x_5,
where $A = 20$, $x_1 = 23$, $x_2 = 12$, $x_3 = 18$, and $x_4 = 22$.

4. Solve each equation for the specified variable in terms of the other variables.

A. $A = bh$ for b

B. $V = (1/3)\pi r^2 h$ for h

C. $A = P + Prt$ for P

D. $P = 2l + 2w$ for w

E. $S = 2lw + 2wh + 2lh$ for l

F. $F = (9/5)C + 32$ for C

5. Suppose $x = m/n$. Write the expression $(m - 3n)/(5n + 2m)$ in terms of x. Assume that $n \neq 0$ and $5n + 2m \neq 0$.

Set 23: Ratios and Proportion

1. Solve each proportion.

A. $x/6 = 7/2$

B. $(v + 6)/7 = (3 - v)/4$

C. $(2x - 7)/11 = (3x + 5)/2$

D. $\dfrac{5x + 2}{7 - x} = \dfrac{1}{3}$

E. $5/(x + 2) = -2/(8x - 1)$

F. $m/15 = 3/5$

G. $(5 + w)/8 = (w + 1)/6$

H. $(4x + 2)/3 = (7x + 2)/9$

I. $\dfrac{9 - 4x}{x + 3} = \dfrac{4}{7}$

J. $12/x = 6/(x - 1)$

2. Solve each word problem.

A. The ratio of the weight of an object on Earth to the weight of the same object on a distant planet is 3 to 8. Bertha weighs 336 lbs on Earth. How much does she weigh on the other planet?

B. On a certain map, 1 cm represents 300 miles. Two towns are 7.2 cm apart on the map. What is the actual distance between the towns?

C. The ratio of the length of a rectangle to its width is 7 to 4. The width of the rectangle is 36 inches. What is the length of the rectangle?

D. Triangles ABC and DEF are similar. The lengths of two sides of $\triangle ABC$ measure 6 cm and 11 cm. The lengths of the corresponding sides of $\triangle DEF$ are 21 ft and x ft. What is x?

E. The ratio of boys to girls in a classroom is 3 to 4. If there are 12 boys in the classroom, then how many girls are there?

F. A worker can unload 20 pallets in half an hour. How many pallets can he unload in 7 hours?

G. The ratio of the number of white people to the number of black people in a certain country is 13 to 2. If the number of black people is 45 million, then how many white people are there?

H. The ratio of the number of yellow people to the number of black people in a certain country is 2 to 5. If the number of yellow people is 18 million, then how many black people are there?

Set 24: Direct and Inverse Variation

1. Write the equation that corresponds to each statement. Let k represent the constant of proportionality.

    A. d is proportional to w D. l is directly proportional to w

    B. P varies inversely as V E. A varies jointly as b and h

    C. V varies directly as r^3 F. c is inversely proportional to d

    G. x varies directly as y and inversely as the square of z

2. Find the constant of proportionality k using the given pieces of information.

    A. P varies directly as s.   B. r varies inversely as t.
       If $s = 7$, then $P = 28$ If $r = 60$, then $t = 4$.

    C. h varies directly as V and inversely as the square of r.
       If $V = 48\pi$ and $r = 4$, then $h = 3$.

    D. S varies jointly as r and h.
       If $r = 2$ and $h = 3$, then $S = 12\pi$.

3. Suppose y varies as x with a constant of proportionality of 3.

    A. Find y when $x = 2$.   B. Find x when $y = 33$.

4. Suppose w varies inversely as l with a constant of proportionality of 24.

    A. Find l when $w = 3$.   B. Find w when $l = 12$.

5. Suppose z varies directly as x and inversely as y.
If $x = 14$ and $y = 2$, then $z = 56$. Find z when $x = 45$ and $y = 5$.

6. Suppose r varies jointly as p and q.
If $p = 5$ and $q = 6$, then $r = 15$. Find p where $q = 8$ and $r = 32$.

7. The cost per person of a cake varies inversely as the number of people who will eat it. The cost is 76 cents per person if fifteen people eat it. How many people must eat it to reduce the cost to 60 cents per person?

8. The weight of a rectangular piece of wood varies jointly as its length, width, and height. If the weight of a piece is 288 when it has length 9, width 8, and height 2, then what is the length of a piece of the same kind of wood that has weight 120, width 4, and height 3?

Set 25: Percentages

1. Convert each percentage to a decimal.

    A. 78%   C. 8%

    B. 350%   D. 0.1%

2. Convert each decimal to a percentage.

    A. 0.35   C. 52.63

    B. 0.00087   D. 0.04

3. Convert each fraction to a percentage.

    A. 3/5   C. 1/50

    B. 7/4   D. 2/3

4. Convert each percentage to a fraction.

    A. 180%   C. 25%

    B. 59%   D. 400%

5. Answer each question.

    A. What is 20% of 55?   G. What is 63% of 200?

    B. What is 150% of 42?   H. What is 1% of 37?

    C. 65% of what number is 39?   I. 12% of what number is 45?

    D. 2% of what number is 9?   J. 250% of what number is 135?

    E. 28 is what percent of 80?   K. 57 is what percent of 75?

    F. 141 is what percent of 100?   L. 3 is what percent of 1000?

6. Solve each word problem.

    A. LaShonda's girth increased from 60 inches to 63 inches in one month. By what percentage did her girth increase?

B. Chuck lost 64% of his savings at the casino. If he lost a total of $3,024, then how much money had he saved before he entered the casino?

C. Priscilla has head lice. On Monday, she had 740 lice. By Friday, the number of lice had increased by 280%. How many lice infested her head on Friday?

D. A tapeworm in Sumil's large intestine grew by 41% between April 1 and June 30. If the tapeworm measured 28.2 cm on June 30, then what was its length on April 1?

E. Emil found a stray dog and gave it a bath, thereby eliminating 99.44% of its fleas. If the dog had 2,500 fleas before the bath, then how many did it have after the bath?

F. While at a museum, Simon chips some ivory off the ancient tusk of an elephant. In doing so, he reduces its weight from 55.0 lbs to 47.3 lbs. By what percentage has the weight of the tusk decreased?

Set 26: Discount and Profit

1. Solve each word problem on discount.

    A. A merchant sells bricks of moldy cheese at a discount of 80% on the listed price of $12 per brick. What is the discounted price?

    B. Nels bought a yoyo at a 18% discount. If the price he paid was $2.87, then what was the regular price of the yoyo?

    C. Tyrone paid only $10.50 for a basketball that normally cost $12.50. What rate of discount (as a percentage of the original price) did he receive on the ball?

    D. A dealer of motor vehicles lowered the price of one of his motorcycles by 5%. If the original price was $2,440, then what is the reduced price?

    E. Sherman reduced the price of one of his tanks by 47%. It now costs only $43,407. What was the price before the reduction?

    F. Bertha lowers the price of her donuts from 80 cents to 52 cents after 2 PM. By what percentage does she lower the price?

2. Solve each word problem on profit.

    A. A bookseller buys a book for $6.20 and sells it at a profit of 115%. At what price did he sell the book?

    B. Jiahong sells fortune cookies for $8.88 per bag. He earns a profit of 85% on each bag. How much does Jiahong pay for each bag?

    C. It cost Domingo $1.20 to produce a taco, which he then sells for $3.30. How much profit does he earns on each taco as a percentage of its cost?

    D. Rose buys a flower for $3 and wants to earn a profit of 78% on its sale. At what price should she sell the flower?

    E. The owner of a service station anticipates a shortage in gasoline and raises the price of his gas by 23%. If the new price is $2.46, then what was the original price?

    F. Esther buys frying pans for $28 and sells them for $40.60. What is her rate of profit as a percentage of her cost?

3. Solve each word problem.

    A. Tatiana presents a coupon to the cashier in a store that enables her to buy a pair of sandals at a 20% discount on its regular price. She then flashes a card that proves her membership in a preferred political party and grants her an extra reduction of 15% on the discounted price. What single discount is equivalent to these two successive discounts?

    B. In one year, the price of an egg rises by 12%. The next year, its price rises by 5%. The third year, its price rises by 30%. By what percentage did the price of the egg rise over this three-year period?

    C. Makoa gains and loses weight seasonally. His weight balloons by 15% in the fall and 25% in the winter as he gobbles poi, but it falls by 10% in the spring and 20% in the summer as he labors in the fields of sugar cane. In exactly one year, by what percentage does his weight change?

Set 27: Simple Interest

1. Solve each word problem.

A. A man deposits $2,000 in an account on which he earns 5% per year in simple interest. How much interest does he earn in eight years?

B. A lender receives $1,350 every quarter on a loan of $20,000 made to a desperate gambler. What annual rate of simple interest does he earn?

C. A girl deposits $100 in a account when she is seven years old. She earns simple interest on the money at an annual rate of 6%. How old will she be at the time she has earned $480 in interest?

D. Tim earns $448 in simple interest every six months on an investment that earns 7% in simple interest per year. What was the initial amount of the investment?

E. Eve earns 4% per year in simple interest on money in an account. If she initially placed $5,000 in the account, then how much interest will she earn over the course of ten years?

F. How should you divide $46,500 between two investments in order to receive $8,992 in interest over 4 years if one investment pays simple interest at an annual rate of 2.2% and the other investment pays simple interest at an annual rate of 5.7%.

G. Elijah invested $150,000 in municipal bonds that pay him $25 every month. What annual rate of simple interest does he earn on the bonds?

H. Ned places a certain amount of money in a account at a bank that pays him simple interest at an annual rate of 2.5%. How long will it take before the amount of interest he has earned is equivalent to the amount of his initial deposit?

I. Zelda receives $200 every month in simple interest on an account that pays 4% annually. How much money did she place in the account?

J. A businessman places money in two funds. In one year, the first fund earns a profit of 12% and the second fund suffers a loss of 7%. If the total amount of money he invested was $10,000 and he suffered a net loss of $301, then how much money did he place in each investment?

K. How long will it take for an investment of $1,000 to quadruple in value if the annual yield is 6% of the original principal?

Set 28: Mixture

1. Solve each word problem.

A. How many liters of pure water should a person add to 12 liters of a solution that is 40% antifreeze to make a solution that is 15% antifreeze?

B. A bartender mixes one pint of a drink that contains 8% alcohol with two pints of a drink that contains 14% alcohol. What is the percentage of alcohol in the final solution?

C. A drum contains 50 gallons of fertilizer that is 14% nitrogen. How many gallons of the fertilizer should a farmer replace with fertilizer that is 30% nitrogen in order to obtain 50 gallons of a fertilizer that is 20% nitrogen?

D. A beaker contains 400 milliliters of solution that is 71% salt. How much water in the beaker must evaporate before the solution that remains is 80% salt?

E. A boy mixes a solution that is 35% salt with a solution that is 60% salt. How many quarts of each solution must he mix to obtain 30 quarts of a solution that is 50% salt?

F. How many grams of an alloy that is 70% nickel should a metallurgist add to 120 grams of an alloy that is 25% nickel to obtain an alloy that is 30% nickel?

G. In a group of 60 people, 30% are brown. In another group of 75 people, 12% are brown. If the two groups are combined, then what percentage of all the people are brown?

H. A swimming pool contains 10,000 gallons of liquid that is 3% chlorine. How many gallons should a swimmer replace with pure water in order to bring the percentage of chlorine in the pool down to 1.2%?

I. A pail contains 8 quarts of a solution that is 42% iodine. If one quart of the water in the pail evaporates, then what is the percentage of iodine in the solution that remains?

J. A hiker mixes old trail mix that is 20% raisins with new trail mix that is 5% raisins. How much of each mix should he combine to obtain a batch of trail mix that is 15% raisins and weighs 48 oz?

Set 29: Adding and Subtracting Polynomials

1. State the degree and leading coefficient of each polynomial.

A. $9x^2 + 4x - 1$

B. -8

C. $13 - 5x^3$

D. $15 + 10x^2 - 6x$

E. x

F. $-2x^3 - 7x^2 + 3x^8 - 9$

G. $-7x^5 - 3x^4 - 7x + 10$

H. $3 + 2x + x^2$

I. 19

J. $6x - 4x^3 + 11x^2$

K. $1 + 2x + 4x^5 + 8x^2$

L. $-x + 4$

2. Simplify each expression.

A. $(7x^2 + x - 9) + (2x^2 - 5x - 2)$

B. $(9x - 2) - (4 - 3x - 6x^2)$

C. $(-7m - 4) + (5x + 2)$

D. $(-2c - 5) - (6c + 9) + (8c + 14)$

E. $(16x - 5) - [9x + 4 - (x + 11)] - 8$

F. $(5x + 1) + (8y - 3)$

G. $(8b^2 + 5) - (-3b + 1)$

H. $(7 + 5x - x^2) + (2 - 3x + 2x^2)$

I. $(-9x^3 + 6x^2 + 2x + 10) - (4x^3 + 5x - 2x + 7)$

J. $(2x^3 - 4x^2 - 7x + 2) + (-11x^2 + 9x - 3)$

K. $(4k^3 - 2k) - (5k^4 - 2k^2 - 17)$

L. $(n^2 + n - 12) + (5n^2 - 3) - (9n^3 + 7n - 8)$

M. $(-w - 6) + [w^3 + 3w^2 - (4w^2 + 2w - 7) + 5w - 3] + 1$

N. $(2a^2 - 2a + 7) + (-3b^2 + 5b + 1)$

Set 30: Properties of Exponents within Products

1. Simplify each expression.

A. $(4x^2)(6x^7)$

J. $(-4p^4q^2)(-7q^{12})$

B. $(-8x^3y)(2x^4y^5)$

K. $(x^6y^3)(-x^3y)(-2xy^3)$

C. $(\frac{3}{2}uv)(4u^3)(-\frac{2}{5}u^2v^7)$

L. $(-3m^2n)^3$

D. $(7x^4)^2$

M. $(3wx^2y^3z^4)^5$

E. $(-a^9b^5c^2)^{10}$

N. $4w^3(1 - 2w - w^2)$

F. $3x(8x + 7)$

O. $-x(1 + x + x^2)$

G. $(3x^4y^7)^2(2xy^2)^5$

P. $mn(m^2 - mn + n^4)$

H. $-10t^7(3t^2 - 4t - 9)$

Q. $(x^5y + x^3y^2 - xy^3)(y^8)$

I. $(9w^5)(3w)$

R. $(2x^5yz)^3(5xy^3z^2)^2$

S. $x(4x^3 - x^2 - 7x) + 5x^2(9x^2 + 3x - 1)$

T. $4x[3x - 4(5x - 2)] - 5(3x^2 + 10x)$

U. $-5x(x^2 - 3x - 13) - (17x^2 + 8x - 2)$

V. $6x^3(3 - 2x) + 4x^2[x^2 + 2(x^2 - 7x)]$

Set 31: Multiplying Polynomials

1. Expand each product.

A. $(x + 2)(x + 8)$

H. $(x + 7)(x - 1)$

B. $(4n - 5)(n + 9)$

I. $(2m - 5)(9m - 3)$

C. $(3b + 2)(6b - 7)$

J. $(x - 10)(3x + 4)$

D. $(c - 2)(d + 3)$

K. $(v + 5)(w + 10)$

E. $(x - 1)(x^2 + x + 1)$

L. $(3x^2 - 7x - 4)(4x^2 + 8x + 3)$

F. $(x^2 + 2)(2x^2 + 3x - 8)$

M. $(y + 3)(y^2 - 3y + 9)$

G. $(m + 2)^3$

N. $(2k - 3)^3$

2. Expand each product using a property for the square of a binomial or the product of the binomials $a + b$ and $a - b$.

A. $(x - 5)^2$

I. $(y + 4)^2$

B. $(3m + 7)^2$

J. $(5x - 9)^2$

C. $(6 - 2m)^2$

K. $(10 - x)^2$

D. $(9x + 4y)^2$

L. $(2a - 7b)^2$

E. $(x + 3)(x - 3)$

M. $(m - 8)(m + 8)$

F. $(7 - 5k)(7 + 5k)$

N. $(4x + 9)(4x - 9)$

G. $(1 + x^2)(1 - x^2)$

O. $(2x^2 + 4)(2x^2 - 4)$

H. $(3m + n)(3m - n)$

P. $(8x - 5y)(8x + 5y)$

3. Suppose $a, b \in \mathbf{R}$.

A. Prove that $(a + b)(a - b) = a^2 - b^2$.

B. Prove that $(a + b)(a^2 - ab + b^2) = a^3 + b^3$.

C. Prove that $(a + b)^4 = a^4 + 4a^3b + 6a^2b^2 + 4ab^3 + b^4$.

Set 32: Multiplying Algebraic Expressions

1. Expand each product.

A. $(3x + 1)(8x + 2)$

E. $(2x - 5)(4x + 9)$

B. $(7a - b - 3c)(2a + 4b + 10c)$

F. $(5x + 4y - 2z)(x - 3y - 8z)$

C. $(x + 2)(x^2 - 4x + 4)$

G. $(c + 3)(c^2 - 5c - 6)$

D. $(2x - 1)(x + 5)(3x - 4)$

H. $(4 - 5x)(8 - x)(7 + 3x)$

2. Determine the degree of each polynomial.

A. $(x^{10} - 2)^5(x^7 + 9)^2$

B. $(3x^3 - 5x^2 - 8x + 1)^7(11 - 7x^2)^4$

Set 33: The FOIL Method

1. Expand each product of binomials using the FOIL method.

A. $(x - 5)(x - 6)$

F. $(x + 1)(x - 10)$

B. $(n + 3)(2n - 9)$

G. $(4m + 1)(m + 8)$

C. $(-3p - 7)(p^2 + 4)$

H. $(4 + 5x)(11x + 2)$

D. $(8 - 5r)(-9 - 4r)$

I. $(6d - 1)(1 - 6d)$

E. $(7c^2 - 3c)(4c^3 + c^2)$

J. $(q^8 - 6)(2q^3 - 5)$

Set 34: Properties of Exponents within Quotients

1. Simplify each expression.

A. x^8/x^3

M. y^2/y^9

B. w/w^{14}

N. m^6/m^6

C. $(5v^4)/(10v^5)$

O. $(-27t^{11})/(3t^4)$

D. $(-72n^9)/(-8n)$

P. $(14x^3)/(-10x^{18})$

E. $(16a^2b^7)/(-4a^{10}b^3)$

Q. $(-5xy^2)/(20x^2y)$

F. $(-u^7v^5w^2)/(3u^5v^5w^6)$

R. $(-10v^7)/(-2v^{10}w^3)$

G. $(-3x^5b^2)/(6x^3)$

S. $(50rs^2t^3)/(-100r^8st^2)$

H. $\dfrac{21x^2y^3 + 49x^8y^4}{-7xy^3}$

T. $\dfrac{-8m^{10}n^7 - 12m^9n^5}{6m^5n^2}$

I. $\dfrac{-3a^2b^6 + 9ab^7}{18a^3b^3}$

U. $\dfrac{10x^4y^6 - 13x^2y}{5x^2y^3}$

J. $(y/5)^2$

V. $(b/3a^2)^3$

K. $(2x^5/y^3)^4$

W. $[a^2/(b^7c^3)]^8$

L. $\left(\dfrac{r^5s^2t^{11}}{r^3s^7t}\right)^9$

X. $\left(\dfrac{54uvw^9}{-6u^5v^3w^2}\right)^2$

Set 35: Dividing Polynomials

1. Divide the polynomials as indicated and explicitly identify the quotient and remainder.

A. $(x^2 - 11x + 15) \div (x - 2)$

G. $(x^2 + 4x + 8) \div (x + 5)$

B. $(2m^2 + 3m - 20) \div (m + 3)$

H. $(-5b^2 - 7b + 12) \div (b - 1)$

C. $(8x^3 + 2x^2 - 4x - 7) \div (4x + 1)$

I. $(6x^2 + 5x + 1) \div (2 - 3x)$

D. $(x^2 - 5x + 7) \div (2 + 2x)$

J. $(x^5 + 2) \div (1 + x)$

E. $(29 - 7v + 4v^2 - v^3) \div (v - 4)$

K. $(4c - 10c^2 + 3c^4) \div (c + 2)$

F. $(y^4 - 81) \div (y - 3)$

L. $(3z^3 - 5z^2 - 10z - 8) \div (2z - 6)$

M. $(27x^3 - 21x^2 - 7x + 1) \div (3x^2 - 2x - 5)$

N. $(15x^3 - 14x^2 + 16x - 32) \div (5x^2 + 2x + 8)$

O. $(7x^4 + 31x^3 - 10x^2 - 67x - 33) \div (7x^2 - 4x - 11)$

P. $(10x^4 + 11x^3 - 6x^2 - 5x - 36) \div (2x^2 - x + 4)$

2. Determine the value of k for which division of the second polynomial by the first polynomial leaves a remainder of c.

A. $x - 3$; $x^2 + 2x + k$; $c = 0$

D. $x + 1$; $2x^2 - 6x + k$; $c = 0$

B. $3x + 5$; $6x^2 + 19x + k$; $c = 0$

E. $2x - 7$; $10x^2 - 37x + k$; $c = 0$

C. $x + 4$; $4x^2 + 13x + k$; $c = 7$

F. $3x - 2$; $21x^2 + x + k$; $c = -4$

3. Determine the values of b and c if division of $x^4 + bx^2 + c$ by $x^2 - 3x + 7$ leaves a remainder of 0.

Set 36: Synthetic Division

1. Synthetically divide the polynomials as indicated and explicitly identify the quotient and remainder.

A. $(x^2 - 3x - 7) \div (x - 3)$

B. $(3b^2 + b + 8) \div (b + 1)$

C. $(x^3 + 64) \div (x + 4)$

D. $(6n^2 + 5n + 3) \div (n - \frac{1}{2})$

E. $(k^3 - 8k^2 + 10k - 3) \div (k - 6)$

F. $(2 - 4y - y^2) \div (y + 5)$

G. $(4x^3 - 27x^2 - 5) \div (x - 7)$

H. $(y^2 + 10y - 5) \div (y + 3)$

I. $(2a^2 - 7a - 11) \div (a - 5)$

J. $(32 - m^5) \div (m - 2)$

K. $(5 - 12x + 3x^2 - 5x^3) \div (x - 1)$

L. $(-12x^2 - 11x - 5) \div (x + \frac{2}{3})$

M. $(v^3 + 9v^2 + 15v + 10) \div (v + 7)$

N. $(3x^3 - 9x - 5) \div (x + 2)$

O. $(x^4 - 3x^3 - 6x^2 - 19x + 15) \div (x - 5)$

P. $(5b^5 + 7b^4 - 4b^2 - b + 11) \div (b + 2)$

Set 37: Nonpositive Exponents

1. Simplify each expression.

A. 8^{-2}

G. 2^{-3}

B. $(11)^0$

H. $1/(7^{-2})$

C. $1/(3^{-4})$

I. $(-4)^0$

D. $(3/2)^{-5}$

J. $(2/5)^{-3}$

E. $[(3^{-1})/7]^{-3}$

K. $[1/(2^{-3} \cdot 8^{-1})]^{-2}$

F. $(6^{-1} \cdot 3^{-2})^{-2}$

L. $[5/(2^{-2})]^4$

2. Simplify each expression. Leave your answer in a form where exponents are positive.

A. $x^{-4}x^{-10}$

I. $y^3 y^{-7}$

B. $(-7m^{-6})(-3m^{11})$

J. $(9n^{-5})(-4n^{-1})$

C. $\dfrac{5w^{-2}}{-30w^{-9}}$

K. $\dfrac{-32x^{-8}}{-18x^{-5}}$

D. $\dfrac{21b^{-6}}{-7b^{13}}$

L. $\dfrac{3m}{33m^{-4}}$

E. $\left(\dfrac{18a^{-8}}{-9a^{-8}}\right)^{-7}$

M. $\left(\dfrac{-12t^{-5}}{3t^3}\right)^{-3}$

F. $\left(\dfrac{-10v^{-3}}{12v^{-7}}\right)^{-2}$

N. $\left(\dfrac{-2n^{-6}}{-8n^{-6}}\right)^{-4}$

G. $\dfrac{x^{-3} \cdot x^{-4}}{x^5 \cdot x^{-12}}$

O. $\dfrac{4(u^{-2})^5 \cdot 3u^3}{6u^8}$

H. $\dfrac{2a^{-4}b^7}{8a^5b^{-2}}$

P. $\dfrac{9r^3s^{-6}}{3r^7s^{-1}}$

Set 38: Factorizing by Removing a Common Factor

1. Determine the prime factorizations and the greatest common factor of the numbers in each set.

A. 18, 32

D. 40, 75

B. 35, 72

E. 56, 71

C. 24, 36, 90

F. 28, 84, 140

2. Determine the greatest common monomial factor of the terms in each set.

A. $30xy$, $48yz$

D. $6y^7$, $2y^5$

B. $108m^7n^2$, $45m^3n^{10}$

E. $33ab^2$, $88ab^4$

C. $12x^2y^9z^5$, $30x^4y^8z^{11}$, $50x^3y^{21}z$

F. $39q^4r^7s^{11}t^9$, $65q^3r^7st^{10}$

3. Factorize each expression completely.

A. $8x^4 + 12x^2$

G. $30y - 40y^8$

B. $9x^2y - 17xy$

H. $57m^8n^3 + 54m^5n^{10}$

C. $4t^5 - 12t^4 + 6t^3$

I. $14u - 35u^2 - 49u^4$

D. $6x(x-2) - 5(x-2)$

J. $m(3c+4) + 2n(3c+4)$

E. $12y^2(7-4z) + 10y(7-4z)$

K. $(2x+10)(x+8) + (x+8)(1-9x)$

F. $(7x-1)(x-2) - (x+4)(x-2)$

L. $20x^5(x^2-3x) - 55x^7(x^2-3x)$

Set 39: Factorizing by Grouping

1. Factorize each expression.

A. $cx + cy + 5x + 5y$

I. $bm - bn + 3m - 3n$

B. $7r^2 - 14s - kr^2 + 2ks$

J. $-u - v - tu - tv$

C. $10m^2 - 5m + 8m - 4$

K. $36y^2 + 24y + 21y + 14$

D. $-18x^3 - 6x^2 + 15x + 5$

L. $12n^3 - 3n^2 - 4n + 1$

E. $100x^3 - 20x^2 + 15x - 3$

M. $-4 - 24x - 8x^2 - 48x^3$

F. $ax + ay - x - y$

N. $4c^4 - 10c^2 + 6c^2 - 15$

G. $9x^6 + 12x^4 - 12x^4 - 16x^2$

O. $8u - tu - 16v + 2tv$

H. $10x - 4y + 6z - 5bx + 2by - 3bz$

P. $7x^4 + 7x^3 - 56x^2 + 5x^2 + 5x - 40$

2. Suppose that $ab = 7$ and $a^2b + a + ab^2 + b = 96$. Evaluate $a^2 + b^2$.
[Note: Do not try to determine the values of a and b.]

Set 40: Factorizing the Difference of Two Squares

1. Factorize each expression completely.

A. $x^2 - 9$

B. $n^2 - 81$

C. $u^2 - v^2$

D. $2w^2 - 72$

E. $4x^2 - 25y^2$

F. $32s^2t^5 - 98s^2t^3$

G. $x^4 - x^2$

H. $4x^3 + 8x^2 - 25x - 50$

I. $m^2 - 64$

J. $x^2 - 121$

K. $x^2 - y^2$

L. $49s^2 - 100t^2$

M. $27u - 12uv^2$

N. $15x^3 - 15x$

O. $3 - 48y^6$

P. $3x^3 - 6x^2 - 27x + 54$

Set 41: Factorizing the Difference or Sum of Two Cubes

1. Factorize each expression completely.

A. $x^3 - 27$

B. $40 + 5b^3$

C. $125 + 64m^3$

D. $y^6 + 343z^3$

E. $8r^3 - 64s^3t^3$

F. $1 - u^9$

G. $x^3 + 8$

H. $1 - 216n^3$

I. $4n + 108m^3n$

J. $a^3b^3 + c^{12}$

K. $x^{12} - 1$

L. $x^6 - 64y^6$

2. Suppose two numbers have a sum of 5 and a product of 3. Determine the sum of their cubes.

Set 42: Factorizing Trinomials of the Form $x^2 + bx + c$

1. Factorize each trinomial.

A. $x^2 + 5x + 6$

B. $y^2 + 8y - 20$

C. $m^2 - 13m + 40$

D. $n^2 - 4n - 60$

E. $c^2 + 9c + 18$

F. $x^2 - 2x - 8$

G. $v^2 - 11v + 24$

H. $w^2 + 4w - 12$

I. $x^2 - x - 56$

J. $b^2 - 17b + 70$

K. $k^2 + 7k + 10$

L. $x^2 + x - 20$

M. $a^2 - 2a - 63$

N. $x^2 + 15x + 54$

O. $z^2 - 3z + 2$

P. $a^2 - 4a - 5$

Q. $b^2 + 4b - 21$

R. $y^2 - 14y + 45$

S. $w^2 + 11w + 28$

T. $v^2 + 2v - 80$

U. $t^2 - 2t - 3$

V. $s^2 + 14s + 40$

W. $r^2 - 11r + 10$

X. $x^2 - 5x - 36$

Y. $c^2 + 4c - 32$

Z. $t^2 - t - 12$

2. Factorize each expression completely.

A. $y^4 + 10y^2 + 24$

B. $3m^2 + 24m - 27$

C. $mn^2 - 8mn + 15m$

D. $5x^2y^3 + 5xy^3 - 30y^3$

E. $x^2 - 18xy + 80y^2$

F. $y^2 - 49$

G. $4m^2 - 100$

H. $a^6 + 2a^3 - 35$

I. $2v^2 - 18v + 28$

J. $b^2c + 15bc + 50c$

K. $3st^2 - 21st - 90s$

L. $m^2 + 6mn - 16n^2$

M. $k^2 - 1$

N. $7c^2 - 28$

47

Set 43: Factorizing Trinomials of the Form $ax^2 + bx + c$

1. Factorize each trinomial completely.

A. $2x^2 + 5x - 12$

B. $3y^2 - 16y + 5$

C. $4a^2 - 12a - 27$

D. $6n^2 + 17n + 10$

E. $6w^2 - 50w + 84$

F. $20k^3 + 60k^2 - 35k$

G. $2x^2 + 25x + 63$

H. $3z^2 - 14z - 24$

I. $4m^2 - 29m + 30$

J. $6b^2 + 13b - 110$

K. $15z^2 - 114z - 48$

L. $36r^2s + 124rs - 160s$

Set 44: Solving Equations by Factorizing

1. Solve each equation.

A. $x^2 + 7x = 0$

B. $2x^2 = 20x$

C. $x^2 = 16$

D. $(x+3)(x+4) = 0$

E. $3x(x+2) - 8(x+2) = 0$

F. $x^2 + 17x + 72 = 0$

G. $x^2 - 6x = 7$

H. $x(x-11) = -30$

I. $x^2 - 5x + 20 = 8x - 22$

J. $x^2 - 12x + 7 = 2x^2 - 3x + 15$

K. $5x^3 = 5x$

L. $(2x-11)(x-2) = -5$

M. $x^3 + 4x^2 + 4x = 0$

N. $x^2 - 3x = 0$

O. $3x^2 = -24x$

P. $49 = x^2$

Q. $x^2 - 12x + 27 = 0$

R. $x^2 + 2x = 48$

S. $(2x-1)(4x+10) = 0$

T. $14(9-2x) - 5x(9-2x) = 0$

U. $x(x+5) = -4$

V. $x^2 + x + 1 = 3x + 9$

W. $4x^2 + 2x + 10 = 5x^2 + 9x - 8$

X. $4x^3 = 36x$

Y. $2x^3 - 14x^2 - 36x = 0$

Z. $(6x+1)(x+1) = 6$

2. Solve each word problem.

A. The length of a table is 2 feet longer that its width. The area of the table is 24 square feet. Find the dimensions of the table.

B. The base of a triangle is 3 inches shorter than its height. The area of the triangle is 35 square inches. Find the dimensions of the triangle.

C. The age of a boy is seven less than twice the age of his sister. The product of their ages is 99. Find the age of the sister.

D. The number of marbles that Larry owns is one more than three times the number of marbles that Abdullah owns. The product of the numbers of marbles that they own is 52. Find the number of marbles that Larry owns.

Set 45: Simplifying Fractions with Variables

1. Simplify each fraction.

A. $\dfrac{21xy}{15x}$

J. $-\dfrac{10a}{25ab}$

B. $\dfrac{-18m^2n^3}{-6mn^5}$

K. $\dfrac{8a^3b^2}{-4a^5b^4}$

C. $\dfrac{-4abc}{6a^2b^5c^3d}$

L. $\dfrac{-9x^4y^8z^3}{-3x^2yz^2}$

D. $\dfrac{2x+8}{6x^2+24x}$

M. $\dfrac{4x^2-20x}{12x-60}$

E. $\dfrac{x^2-3x-4}{x^2+7x+6}$

N. $\dfrac{x^2-7x+12}{x^2-16}$

F. $\dfrac{2x^2+11x-40}{2x^2-9x+10}$

O. $\dfrac{3x^2-23x+14}{5x^2-29x-42}$

G. $\dfrac{9x^2+30x+25}{3x^2+2x-5}$

P. $\dfrac{6x^2-29x+30}{14x^2-19x-3}$

H. $\dfrac{4x^2-25}{15x-6x^2}$

Q. $\dfrac{3x^2-11x-20}{45+x-2x^2}$

I. $\dfrac{5x^2-28x+32}{x^3-64}$

R. $\dfrac{8x^3+27}{2x^2-19x-33}$

Set 46: Multiplying and Dividing Fractions with Variables

1. Perform the indicated operations and simplify the resulting fractions.

A. $\dfrac{4}{3} \cdot \dfrac{15}{2}$

F. $\dfrac{9}{7} \div \dfrac{12}{35}$

B. $-\dfrac{3xy}{10z} \div \dfrac{3x}{2z}$

G. $\left(-\dfrac{6a^2}{5b^3c}\right)\left(-\dfrac{20b^2c^5}{3a^6}\right)$

C. $\dfrac{8x+24}{5xy-20y} \cdot \dfrac{15x^2y-60xy}{3x+9}$

H. $\dfrac{mn+9m}{2m-12} \div \dfrac{n^2+11n+18}{7m^2-7m-210}$

D. $\dfrac{x^2+3x}{x^2-9x+20} \div \dfrac{x^3-x^2-12x}{2x^2-9x-5}$

I. $\dfrac{2x^2-18x+28}{3x^2+19x-40} \cdot \dfrac{x^2+9x+8}{4x^2-20x-56}$

E. $\dfrac{b^2-25}{b^2+13b+40} \cdot \dfrac{b^2+6b-16}{b^2-7b+10}$

J. $\dfrac{c^2+5c-14}{c^2-4c-12} \div \dfrac{2c^2-13c+18}{c^2+c-42}$

Set 47: Adding and Subtracting Fractions with Variables

1. Perform the indicated operations and simplify the resulting fractions.

A. $\dfrac{2x-3}{6} + \dfrac{5x+4}{6}$

L. $\dfrac{2-9m}{5} - \dfrac{8m}{5}$

B. $\dfrac{5n}{6} + \dfrac{3n-8}{10}$

M. $-\dfrac{x+11}{12} + \dfrac{7x-1}{3}$

C. $\dfrac{2(p+6)}{5p^2} - \dfrac{4(3-7p)}{5p^2}$

N. $\dfrac{5(4y+5)}{16y+22} + \dfrac{4(y+2)}{16y+22}$

D. $\dfrac{8}{x+1} + \dfrac{3}{x-5}$

O. $\dfrac{1}{2x-3} - \dfrac{4}{x}$

E. $\dfrac{5x+3}{4x^2} - \dfrac{9}{14x}$

P. $\dfrac{2x-9}{8x} + \dfrac{x^2-3}{2x^2}$

F. $\dfrac{10x-3}{4x} + \dfrac{5}{2x-2}$

Q. $\dfrac{x+7}{9x-6} - \dfrac{3x-5}{6x-4}$

G. $\dfrac{5-m}{10n} - \dfrac{2+3n}{15m}$

R. $\dfrac{b-12c+2}{8bc} + \dfrac{3a+2b-6}{2ab}$

H. $\dfrac{8}{x^2+3x-10} + \dfrac{5}{x^2+5x}$

S. $\dfrac{x-2}{x^2+6x-7} - \dfrac{x+1}{x^2-5x+4}$

I. $\dfrac{-3x+4}{x^2-12x+20} - \dfrac{9x-5}{x^2-x-2}$

T. $\dfrac{6x}{x^2+10x+21} + \dfrac{4-3x}{x^2-2x-15}$

J. $\dfrac{x-11}{2x^2-3x-20} + \dfrac{4x+7}{2x^2-11x-40}$

U. $\dfrac{4x}{6x^2-17x+7} - \dfrac{x+1}{3x^2+20x-63}$

K. $\dfrac{4-n}{n^3+8} - \dfrac{7}{3n^2+19n+26}$

V. $-\dfrac{9+m}{m^3-27} + \dfrac{5}{2m^2-11m+15}$

W. $\dfrac{5}{x^2-4} - \dfrac{1}{x^2-2x} - \dfrac{4}{x^2+2x}$

X. $\dfrac{4x-3}{x^2-4x-5} + \dfrac{7x+2}{x^2+7x+6} - \dfrac{2-9x}{x^2+x-30}$

Set 48: Complex Fractions

1. Simplify each complex fraction.

A. $\dfrac{4/(3x)}{10/(21x)}$

I. $\dfrac{6x^2/11}{3x/22}$

B. $\dfrac{5/4 - 2/7}{9/2 + 1/5}$

J. $\dfrac{4/5 + 7/2}{2/3 + 1/9}$

C. $\dfrac{-x^{-5} - y^{-2}}{x^{-3} - y^{-1}}$

K. $\dfrac{x^{-1} + y^{-3}}{x^{-7} - y^{-4}}$

D. $\dfrac{\dfrac{x^2 + 4x - 5}{x + 2}}{\dfrac{3x - 3}{x^2 + 9x + 14}}$

L. $\dfrac{\dfrac{x^2 - 25}{x^2 - 10x + 16}}{\dfrac{2x + 10}{9x - 18}}$

E. $\dfrac{\dfrac{x}{4} - \dfrac{1}{x}}{\dfrac{1}{5} - \dfrac{6}{15x}}$

M. $\dfrac{\dfrac{7}{2x} + \dfrac{4}{3}}{\dfrac{16}{7x} + \dfrac{6}{x^2}}$

F. $\dfrac{\dfrac{3}{x + h + 2} - \dfrac{3}{x + 2}}{h}$

N. $\dfrac{\dfrac{8}{(x + h)^2} - \dfrac{8}{x^2}}{h}$

G. $\dfrac{\dfrac{8}{x - 2} - \dfrac{4}{x + 4}}{\dfrac{3}{x + 1} - \dfrac{1}{x + 7}}$

O. $\dfrac{\dfrac{4}{x} + \dfrac{2}{x + 6}}{\dfrac{3}{x + 2} - \dfrac{6}{x}}$

H. $2 - \dfrac{3}{2 - \dfrac{3}{2 - \dfrac{3}{x + 4}}}$

P. $2 + \dfrac{15}{2 + \dfrac{15}{2 + \dfrac{15}{2 + \dfrac{15}{2 + \dfrac{15}{\vdots}}}}}$

Set 49: Equations with Fractions

1. Solve each equation.

A. $\dfrac{x}{2} - \dfrac{3x}{5} = \dfrac{3}{4}$

I. $\dfrac{7x}{8} + \dfrac{x+2}{3} = 4$

B. $\dfrac{1}{u} + \dfrac{11}{4} = \dfrac{1}{2}$

J. $\dfrac{3}{7w} - \dfrac{1}{3w} = \dfrac{5}{3}$

C. $\dfrac{5}{4x-2} - \dfrac{9}{2x-1} = \dfrac{4}{11}$

K. $-\dfrac{12}{x+3} - \dfrac{5}{2} = \dfrac{7}{4x+12}$

D. $\dfrac{2x}{x+8} + 7 = \dfrac{9x}{x-3}$

L. $\dfrac{6x}{x+1} - \dfrac{x}{x-9} = 5$

E. $\dfrac{3x}{x+4} + \dfrac{x-6}{x} = \dfrac{48}{x^2+4x}$

M. $3 - \dfrac{8}{x^2-5x+6} = \dfrac{2x}{x-3}$

F. $\dfrac{x}{x-4} - \dfrac{12}{x^2-2x-8} = \dfrac{2}{x+2}$

N. $\dfrac{2}{x+8} - \dfrac{15}{x^2+11x+24} = \dfrac{x}{x+3}$

G. $\dfrac{6x}{x-7} - \dfrac{5x}{x-2} = \dfrac{22x}{x^2-9x+14}$

O. $\dfrac{x}{x+3} + \dfrac{3x+10}{x^2+3x} = 2$

H. $\dfrac{x}{x-4} - \dfrac{4}{x+5} = \dfrac{36}{x^2+x-20}$

P. $\dfrac{5x}{x+2} + \dfrac{4x-2}{x^2+3x+2} = -\dfrac{6}{x+1}$

2. Solve each word problem.

A. The sum of 5 and the quotient of 4 and the sum of a number and 8 equals the quotient of -36 and the sum of the number and 3. Find the number.

B. The quotient of 15 and the sum of a number and 2 equals the difference of 7 and the quotient of 12 and the number. Determine the number.

Set 50: Rates of Travel

1. Solve each word problem.

   A. A bus moves at a speed of 42 mi/hr after departing a depot. Three hours later, a second bus departs the depot but moves at a speed of 48 mi/hr. How long will it take for the second bus to catch the first bus if they follow the same route?

   B. Two trains leave a station at the same time. One train moves 7 km/hr faster than the other train. If the two trains move in the same direction on parallel tracks, then how far apart will the trains be in five hours?

   C. Two trains leave a station at the same time. One train moves 8 km/hr faster than the other train. If the two trains move in opposite directions, then what are the average speeds of the trains if they are 518 km apart after seven hours?

   D. Kwame runs past a sleeping cheetah at a speed of 23 m/s. Six seconds later, the roused cheetah begins to chase Kwame at a speed of 27 m/s. How long will it take for the cheetah to catch Kwame?

   E. A woman jogs to a restaurant at a rate of 6 mi/hr to gorge herself on pizza. Several hours later, she waddles home along the same route at a speed of 2 mi/hr. Find the distance between her home and the restaurant if the total time she traveled to and from the restaurant was 20 minutes.

   F. Eli travels ten miles to a park to play hooky. He first rides a skateboard downhill at an average speed of 18 mi/hr, but he then must carry the skateboard uphill at a speed of 2 mi/hr. If it takes Eli three hours to get to the park, then how far was he able to ride the skateboard?

   G. An actor demonstrates to reporters his aversion to petroleum by riding his bicycle two kilometers from his home to a parking lot. Having escaped the press, he finds his truck in the lot and drives twelve more kilometers to the studio. If his speed in the truck was 16 km/hr greater than his speed on the bicycle and he traveled the entire distance in 45 minutes, then find his rate of speed in the truck.

   H. Two men participate in a relay race. The first man runs 5 miles before passing the baton to the second man, who runs another 5 miles. If the second man ran 4 mi/hr slower than the first man and together they finished the relay in 1 hour 20 minutes, then find the rate of speed of the slower runner.

Set 51: Rates of Work

1. Solve each word problem.

   A. Enrique can make 20 pizzas in one hour, while Fredo can make 30 pizzas in one hour. If they work together, how long would it take them to make 75 pizzas?

   B. Quentin can sweep the floor of a prison in 80 minutes. His cellmate Ronald can sweep the same floor in 25 minutes. If they work together, how long would it take them to sweep the floor?

   C. The water from both faucets in a sink can fill the sink in 11 minutes. Water from the hot faucet alone can fill the sink in 23 minutes. How long would it take water from the cold faucet alone to fill the sink?

   D. Working together, a robot and a human can install ornaments on the hoods of 264 cars in 45 minutes. Working alone, the human can install only 10% of the number of ornaments that the robot can install by itself. How many ornaments can the human install in one hour?

   E. Sadie can change 12 diapers in ten minutes, whereas Tara can change only 9 diapers in ten minutes. How long would it take them to change 84 diapers if they worked together?

   F. Swift Horse can make 13 more arrowheads per week than Wide Sloth. Working together, they can make 285 arrowheads in three weeks. How many arrowheads can Wide Sloth make per week?

   G. Frank can eat 14 fewer hot dogs in an hour than Joey can. How many hot dogs can Frank eat in one hour if together they can eat 48 hot dogs in forty minutes?

   H. Gus can spade a garden in 35 minutes. Hank can spade the same garden in 40 minutes. How long would it take them to spade the garden if they worked together?

   I. Martha can darn 45 pairs of socks in three fewer hours than Nancy can. How long does it take Martha to darn 45 pairs of socks if together they take five hours to darn 80 pairs of socks?

   J. Ken requires two more hours to produce 77 widgets than Lloyd does. How long does it take Ken to produce 77 widgets if together they take three hours to produce 108 widgets?

K. Gabby and Khawla, the two wives of Masood, must scrub a large tiled floor in their villa every day. Gabby will scrub the floor in six hours when working alone, while Khawla will scrub the floor in seven hours by herself. But when they work together, their combined rate of work falls by twenty-five tiles per hour because they waste time in sharing malicious gossip. If it takes them exactly four hours to scrub the floor when working together, then how many tiles does the floor have?

L. Mort could build a certain brick wall in eleven hours. Red would need only eight hours to build the same wall. But when they work together, their synergy raises their joint rate of production by 42 bricks per hour. If these masons take exactly four hours to build the wall when working together, then how many bricks does the wall have?

Set 52: Sign Charts

1. Construct a sign chart for each expression.

A. $(x+1)(x+5)$

I. $(x-7)(x+4)$

B. $(2x+5)(5x-4)$

J. $(7x-1)(8x-3)$

C. $(4-x)(5-3x)$

K. $(11-2x)(10-x)$

D. $x^2-6x-27$

L. $x^2-10x+16$

E. $3(x-5)(x-7)$

M. $-4(2x+1)(x+8)$

F. $x(x-4)^2(x+8)^3$

N. $-9x^2(x+1)^5(x-99)^{40}$

G. $\dfrac{3x-5}{x-12}$

O. $\dfrac{x+2}{8x+1}$

H. $\dfrac{-8x(4-x)}{(2x-9)^3(x+6)^4}$

P. $\dfrac{-x^3(x+3)^2}{(3x+7)^5(1-6x)}$

2. Use the method of test points to construct a sign chart for each expression given the zeros of its factors.

A. expression: $60x^3-131x^2-27x+14$; zeros: $-\frac{2}{5}, \frac{1}{4}, \frac{7}{3}$

B. expression: $-112x^4+50x^3+29x^2-12x$; zeros: $-\frac{1}{2}, 0, \frac{3}{8}, \frac{4}{7}$

C. expression: $2x^3+3x^2-10x-15$; zeros: $-\sqrt{5}, -\frac{3}{2}, \sqrt{5}$
[Hint: $\sqrt{5} \approx 2.236$]

D. expression: $-x^3+8x^2+7x-56$; zeros: $-\sqrt{7}, \sqrt{7}, 8$
[Hint: $\sqrt{7} \approx 2.646$]

Set 53: Inequalities with Polynomials or Fractions

1. Solve each inequality using a sign chart.

A. $x^2 - 6x - 7 \geq 0$

N. $x^2 + 7x + 12 \leq 0$

B. $2x^2 - 15x + 25 < 0$

O. $3x^2 + 5x - 12 > 0$

C. $x^3 + 13x^2 + 30x > 0$

P. $x^3 - 6x^2 - 72x \geq 0$

D. $5x^3 + 2x^2 - 45x - 18 \leq 0$

Q. $21x^3 - 3x^2 - 84x + 12 < 0$

E. $x^2 + 8x > 20$

R. $x^2 - 10x + 15 \leq x - 15$

F. $\dfrac{2x + 3}{x - 8} \geq 0$

S. $\dfrac{x + 15}{3x + 9} < 0$

G. $\dfrac{x^2 - 3x + 2}{x} \leq 0$

T. $\dfrac{x - 4}{x^2 + 2x - 35} > 0$

H. $\dfrac{5x^2 + 16x + 3}{3x^2 - 14x + 16} < 0$

U. $\dfrac{2x^2 - 11x - 63}{x^2 + 12x + 11} \geq 0$

I. $\dfrac{x^2 + 4x}{x^2 - x - 20} \geq 0$

V. $\dfrac{x^2 - 9x + 18}{x^2 + 5x - 24} < 0$

J. $\dfrac{7}{4x + 8} < 3$

W. $\dfrac{x - 11}{x + 5} \leq -4$

K. $\dfrac{x}{x - 4} > -\dfrac{4}{x + 2}$

X. $\dfrac{2x}{x - 3} \geq \dfrac{5}{x - 4}$

L. $\dfrac{6 + 16x - 6x^2}{10 + 7x + x^2} \leq 0$

Y. $\dfrac{28 - 3x - x^2}{72 - 7x - 2x^2} < 0$

M. $\left| \dfrac{x - 3}{x + 5} \right| \geq 2$

Z. $\left| \dfrac{3x + 1}{x - 2} \right| \leq 4$

Set 54: Roots and Radicals

1. Evaluate each radical without a calculator.

A. $\sqrt{49}$

B. $-\sqrt{16}$

C. $\sqrt[3]{64}$

D. $\sqrt[3]{-1}$

E. $\sqrt{36/81}$

F. $\sqrt[4]{16}$

G. $\sqrt[5]{1}$

H. $\sqrt{0.04}$

I. $\sqrt{-144}$

J. $\sqrt{9}$

K. $-\sqrt{64}$

L. $\sqrt[3]{27}$

M. $\sqrt[3]{-8}$

N. $\sqrt{121/25}$

O. $\sqrt[4]{81}$

P. $\sqrt[5]{-32}$

Q. $\sqrt{0.25}$

R. $\sqrt[8]{-7}$

2. Simplify each expression.

A. $2\sqrt{3}+5\sqrt{3}$

B. $4\sqrt[3]{6}-9\sqrt[3]{6}$

C. $-6\sqrt{5}+13\sqrt{5}-4\sqrt{5}$

D. $\sqrt[4]{17}-\sqrt{17}-5\sqrt{17}-8\sqrt[4]{17}$

E. $5\sqrt{2}+8\sqrt{5}-\sqrt{5}-10\sqrt{2}$

F. $11\sqrt{2}-3\sqrt{2}$

G. $-8\sqrt[4]{7}+\sqrt[4]{7}$

H. $4\sqrt{11}+7\sqrt{11}+3\sqrt{11}$

I. $2\sqrt[3]{13}-7\sqrt{13}-13\sqrt[3]{13}-\sqrt{13}$

J. $9\sqrt{3}+7\sqrt{3}+2\sqrt{10}-3\sqrt{10}$

Set 55: Properties of Radicals

1. Evaluate each radical.

A. $\sqrt{(7)^2}$

G. $\sqrt{(-3)^2}$

B. $\sqrt{(-39)^2}$

H. $\sqrt{(211)^2}$

C. $\sqrt[3]{(-5)^3}$

I. $\sqrt[3]{(6)^3}$

D. $\sqrt[4]{(10)^4}$

J. $\sqrt[6]{(-23)^6}$

E. $\sqrt[20]{(-13)^{20}}$

K. $\sqrt[16]{(71)^{16}}$

F. $\sqrt[7]{(2)^7}$

L. $\sqrt[5]{(-5)^5}$

2. Simplify each expression.

A. $\sqrt{16 \cdot 3}$

I. $\sqrt{25 \cdot 7}$

B. $\sqrt{9 \cdot 6}$

J. $\sqrt{4 \cdot 5}$

C. $\sqrt[3]{8 \cdot 10}$

K. $\sqrt[3]{-27 \cdot 4}$

D. $\sqrt[3]{-64 \cdot 3}$

L. $\sqrt[3]{125 \cdot 2}$

E. $\sqrt{5/81}$

M. $\sqrt{11/36}$

F. $\sqrt{3/49}$

N. $\sqrt{7/64}$

G. $\sqrt[3]{-2/27}$

O. $\sqrt[4]{5/81}$

H. $\sqrt[5]{13/32}$

P. $\sqrt[3]{-17/8}$

3. Simplify the expression $|x - \sqrt{(x-5)^2}|$ where $x < 0$.

4. Determine whether each expression is positive or negative.

A. $10 - 3\sqrt{11}$

B. $7 - 5\sqrt{2}$

Set 56: Simplest Radical Form

1. Convert each radical into simplest radical form. Assume that all variables represent positive real numbers.

A. $\sqrt{12}$

J. $\sqrt{98}$

B. $\sqrt{54}$

K. $\sqrt{500}$

C. $\sqrt{6480}$

L. $\sqrt{3675}$

D. $\sqrt[3]{56}$

M. $\sqrt[3]{-81}$

E. $\sqrt{x^5}$

N. $\sqrt{m^{11}}$

F. $\sqrt[3]{n^{17}}$

O. $\sqrt[6]{c^{23}}$

G. $\sqrt[8]{k^{25}}$

P. $\sqrt[5]{t^6}$

H. $\sqrt{75m^{15}}$

Q. $\sqrt{360x^3}$

I. $\sqrt[3]{-54x^4y^6z^{11}}$

R. $\sqrt[4]{80x^{12}y^7z^{21}}$

2. Convert each expression into simplest radical form. Assume that all variables represent positive real numbers.

A. $\dfrac{3}{\sqrt{10}}$

G. $\dfrac{5}{\sqrt{3}}$

B. $\dfrac{2}{\sqrt[3]{9}}$

H. $\dfrac{11}{\sqrt[3]{2}}$

C. $\dfrac{20x^2}{\sqrt{5x^7}}$

I. $\dfrac{14mn^9}{\sqrt{7m^5n^{11}}}$

D. $\sqrt{\dfrac{7}{20}}$

J. $\dfrac{3}{\sqrt[5]{k^2m^{14}}}$

E. $\dfrac{5}{\sqrt[6]{bc^{13}}}$

K. $\sqrt{\dfrac{5}{18}}$

F. $\sqrt[3]{\dfrac{m^4}{n^{11}}}$

L. $\sqrt[7]{\dfrac{m^{17}}{n}}$

3. Simplify each expression.

A. $5\sqrt{2} - 3\sqrt{8}$

B. $-\sqrt{20} + \sqrt{125} - \sqrt{5}$

C. $\dfrac{18\sqrt{12}}{5\sqrt{27}}$

D. $\dfrac{7\sqrt{200}}{3\sqrt{72}}$

E. $\sqrt{27} + 6\sqrt{12}$

F. $\dfrac{\sqrt{6}}{2\sqrt{54}}$

G. $\dfrac{20\sqrt{28}}{4\sqrt{175}}$

H. $-4\sqrt{90} - 7\sqrt{40} - 3\sqrt{10}$

Set 57: Products of Expressions with Radicals

1. Simplify each product. Assume that all variables represent positive real numbers.

A. $\sqrt{2}\sqrt{8}$

B. $\sqrt{18}\sqrt{5}$

C. $(7\sqrt{6})(3\sqrt{10})$

D. $9\sqrt{5}(-\sqrt{10} - 7\sqrt{15})$

E. $(\sqrt{2} + \sqrt{7})(\sqrt{14} - \sqrt{21})$

F. $(4\sqrt{12} + 9\sqrt{3})(2\sqrt{6} + 10\sqrt{15})$

G. $\sqrt[3]{2}\sqrt[3]{16}$

H. $\sqrt{a^3b}\sqrt{a^5b^9}$

I. $(7 - \sqrt{x})(3 - 2\sqrt{x})$

J. $\sqrt{5}(\sqrt[3]{10} - \sqrt{10})$

K. $\sqrt{12}\sqrt{3}$

L. $\sqrt{2}\sqrt{11}$

M. $(2\sqrt{14})(-4\sqrt{7})$

N. $\sqrt{3}(5\sqrt{6} - 4\sqrt{3})$

O. $(\sqrt{12} - \sqrt{11})(\sqrt{2} - \sqrt{5})$

P. $(8\sqrt{10} - 3\sqrt{13})(8\sqrt{10} + 3\sqrt{13})$

Q. $\sqrt[3]{9}\sqrt[3]{3}$

R. $\sqrt{k^4n^5}\sqrt{kn^2}$

S. $(2 + 3\sqrt{m^7})(4 - 6\sqrt{m^7})$

T. $\sqrt[3]{4}(\sqrt{8} + \sqrt[3]{12})$

Set 58: Quotients of Expressions with Radicals

1. Simplify each expression. Assume that all variables represent positive real numbers.

A. $\sqrt{3/5}$

B. $\sqrt{17}/\sqrt{6}$

C. $\sqrt{3}/(2\sqrt{15})$

D. $\sqrt{7}/\sqrt{63}$

E. $4\sqrt{50}/\sqrt{2}$

F. $10/\sqrt{3m}$

G. $9x^5/\sqrt{x^3}$

H. $\sqrt[3]{32}/\sqrt[3]{c^7}$

I. $3\sqrt{10} - 8\sqrt{2/5}$

J. $\dfrac{2}{5+\sqrt{3}}$

K. $\dfrac{-8}{3-\sqrt{2}}$

L. $\dfrac{\sqrt{7}}{6-2\sqrt{7}}$

M. $\dfrac{3-\sqrt{x}}{8+\sqrt{x}}$

N. $\sqrt{7/2}$

O. $\sqrt{4}/\sqrt{10}$

P. $\sqrt{2}/(5\sqrt{14})$

Q. $\sqrt{5}/\sqrt{20}$

R. $11\sqrt{54}/\sqrt{6}$

S. $6v/\sqrt{5w}$

T. $2x/\sqrt{x^7}$

U. $\sqrt[5]{m^9/n^{17}}$

V. $2\sqrt{7} + 3/\sqrt{7}$

W. $\dfrac{20}{2-\sqrt{9}}$

X. $\dfrac{4}{7+3\sqrt{5}}$

Y. $\dfrac{10-\sqrt{3}}{10+\sqrt{3}}$

Z. $\dfrac{2\sqrt{n}+5}{9\sqrt{n}-4}$

2. Rationalize the numerator of each expression.

A. $\dfrac{8+3\sqrt{11}}{7}$

B. $\dfrac{\sqrt{15}-\sqrt{11}}{12}$

C. $\dfrac{5\sqrt{12}+10\sqrt{3}}{11}$

D. $\dfrac{9-4\sqrt{5}}{13}$

E. $\dfrac{2\sqrt{3}+3\sqrt{2}}{9}$

F. $\dfrac{6\sqrt{2}-2\sqrt{18}}{5}$

65

Set 59: Equations with Radicals

1. Solve each equation by squaring both sides.

A. $\sqrt{x-5}=2$

J. $\sqrt{7x+2}=3$

B. $\sqrt{8-3x}=-10$

K. $\sqrt{10+x}=-7$

C. $|m+4|=1$

L. $|2n-9|=5$

D. $|10-3k|=8$

M. $|1+h|=9$

E. $\sqrt{3x+11}=\sqrt{5x-5}$

N. $\sqrt{7x-12}=\sqrt{10x+3}$

F. $3\sqrt{x+6}=8$

O. $7\sqrt{25-6x}=28$

G. $\sqrt{x+2}=x-4$

P. $\sqrt{5x+1}=7-x$

H. $\sqrt{x^2-2x-8}=10-x$

Q. $\sqrt{x^2+4x-17}=x+1$

I. $\sqrt{x^2+3x+9}=2x-3$

R. $\sqrt{2x^2-4x+11}=2x+1$

2. Solve each equation by squaring both sides twice.

A. $\sqrt{3x+13}=2+\sqrt{2x+1}$

C. $\sqrt{x+6}=-1+\sqrt{3x-5}$

B. $\sqrt{4x+1}-\sqrt{x-2}=3$

D. $\sqrt{7x+4}+\sqrt{2x+3}=2$

3. Solve each equation.

A. $\sqrt[5]{3x+11}=2$

E. $\sqrt[4]{1-5x}=3$

B. $\sqrt[3]{n^2-n+15}=5$

F. $\sqrt[3]{m^2-13m-28}=-4$

C. $x=\sqrt[3]{45+9x-5x^2}$

G. $w=\sqrt[3]{3w^2+16w-48}$

D. $\sqrt[3]{v^3+7v^2+3v+16}=v+2$

H. $\sqrt[3]{x^3-4x^2-2x-7}=x-3$

66

Set 60: Equations with Constant Exponents

1. Solve each equation.

A. $x^2 = 9$ K. $x^2 = 36$

B. $x^3 = 125$ L. $x^3 = -27$

C. $x^4 = 256$ M. $x^5 = 243$

D. $x^7 = -128$ N. $x^6 = 1$

E. $x^2 + 5 = 29$ O. $x^2 - 10 = 88$

F. $4x^2 - 81 = 0$ P. $25x^2 + 40 = 161$

G. $(x - 5)^2 = 16$ Q. $(x + 11)^2 = 4$

H. $(3x + 1)^2 - 12 = 13$ R. $(4x - 7)^2 - 23 = 77$

I. $5(x - 3)^4 - 9 = 71$ S. $2(x + 8)^3 + 11 = 65$

J. $\dfrac{(x + 2)^2}{3} = 48$ T. $\dfrac{(x - 6)^2 - 1}{4} = 20$

2. Solve each equation.

A. $x^{x^{x^{x^{x^{\cdot^{\cdot^{\cdot}}}}}}} = 7$ B. $x^{x^{x^{x^{x^{\cdot^{\cdot^{\cdot}}}}}}} = \frac{1}{2}$ [Hint: $x^{1/n} = \sqrt[n]{x}$]

3. Simplify each expression.

A. $\sqrt{7 + 3\sqrt{5}} - \sqrt{7 - 3\sqrt{5}}$ B. $\dfrac{\sqrt{6} + \sqrt{10}}{\sqrt{4} + \sqrt{15}}$

C. $\dfrac{2\sqrt{10}}{\sqrt{2} + \sqrt{5} + \sqrt{7}}$ D. $\dfrac{2\sqrt{30}}{\sqrt{5} - \sqrt{6} - \sqrt{11}}$

E. $3x + \sqrt{9x^2 + 2} - \dfrac{2}{3x + \sqrt{9x^2 + 2}}$ F. $2n - \sqrt{4n^2 + 7} - \dfrac{7}{2n - \sqrt{4n^2 + 7}}$

Set 61: The Pythagorean Theorem

1. Determine the value of the indicated variable. Assume that a and b represent the lengths of the legs of a right triangle and c represents the length of its hypotenuse.

   A. Find b if $a = 5$ and $c = 13$.   D. Find a if $b = 4$ and $c = 5$.

   B. Find a if $b = 3$ and $c = 9$.   E. Find b if $a = 2$ and $c = 7$.

   C. Find c if $a = 6$ and $b = 8$.   F. Find c if $a = 8$ and $b = 15$.

2. Solve each word problem.

   A. One leg of a right triangle is two feet shorter than the other leg. The hypotenuse is two feet longer than the longer leg. Find the lengths of all three sides of the triangle.

   B. The hypotenuse of a right triangle is 1 cm longer than its longer leg, while the shorter leg is 4 cm less than the average of the lengths of the longer leg and the hypotenuse. Find the lengths of all three sides of the triangle.

   C. An eleven-foot ladder is leaning against a wall. The distance from the base of the ladder to the wall is four feet. Determine the distance from the floor to the top of the ladder.

   D. A man walks 90 meters north on a road from point A. He then turns and walks 120 meters west on another road to point B. If the man had walked directly from point A to point B across an open field, how many fewer meters would he have walked?

   E. Simon walks his pet snake on a tight leash. The snake slithers 3 meters ahead of Simon when he holds the leash 1.25 meters above the ground. How long is the leash?

   F. The distance between the lower left and upper right corners of a rectangular flag is 35 inches. The width of the flag is 18 inches. Find the length of the flag.

   G. A vendor places a spherical scoop of ice cream atop an inverted sugar cone with a rim (that is, a base) that measures 6 cm in diameter. The bottom of the scoop of ice cream lies 2 cm below the rim of the cone. Determine the radius of the scoop.

3. Show that the distance d between the lower left front corner and the upper right rear corner of a box with length l, width w, and height h is given by the formula $d = \sqrt{l^2 + w^2 + h^2}$.

4. Determine the area of a square inscribed in a circle of circumference 12 inches (see figure at right).

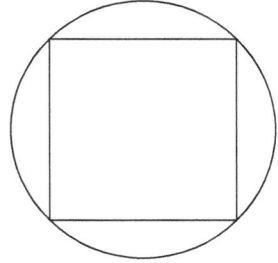

Set 62: Completing the Square

1. Complete the square on each expression.

A. $x^2 - 10x - 3$ I. $x^2 + 6x + 4$

B. $v^2 + 4w - 12$ J. $w^2 - 2w + 9$

C. $x^2 + 5x + 8$ K. $x^2 - 9x - 1$

D. $m^2 - 7x + 5$ L. $n^2 + n - 10$

E. $3x^2 - 12x - 7$ M. $2x^2 + 12x - 5$

F. $4x^2 + 8x + 3$ N. $5x^2 - 10x + 2$

G. $2x^2 + 3x - 6$ O. $2x^2 + 10x$

H. $3x^2 - 15x$ P. $3x^2 - 5x - 6$

2. Solve each equation by completing the square.

A. $x^2 + 8x - 9 = 0$ K. $x^2 - 12x - 13 = 0$

B. $h^2 - 6h = -5$ L. $k^2 + 4k = 21$

C. $x(x - 1) = 30$ M. $x(x + 3) = 10$

D. $x^2 = 99 - 2x$ N. $x^2 = 10x + 39$

E. $2x^2 + 6x = 20$ O. $2x^2 = 2x + 18$

F. $3x^2 = 18x - 24$ P. $5x^2 = -10x$

G. $2x^2 - 5x + 3 = 0$ Q. $2x^2 + 11x + 5 = 0$

H. $3x^2 + 7x = 1$ R. $4x^2 = 9x - 3$

I. $5x^2 + 8x + 3 = 0$ S. $3x^2 = 10x - 3$

J. $2x^2 = 4x - 5$ T. $2x^2 + 7x + 9 = 0$

Set 63: The Quadratic Formula

1. Use the quadratic formula to solve each equation.

A. $x^2 + 8x + 15 = 0$

J. $x^2 - 4x - 117 = 0$

B. $v^2 + 10v = 11$

K. $w^2 - 9w = -14$

C. $m^2 = 5m + 10$

L. $n^2 = 3 - 7n$

D. $k^2 - k + 3 = 0$

M. $h^2 + 6h + 10 = 0$

E. $3x^2 - 17x + 10 = 0$

N. $2x^2 - x - 3 = 0$

F. $5x^2 + 12x + 5 = 0$

O. $4x^2 + 8x - 1 = 0$

G. $\frac{1}{4}x^2 - \frac{1}{2}x + \frac{1}{4} = 0$

P. $\frac{1}{9}x^2 + \frac{2}{3}x + 1 = 0$

H. $\frac{1}{8}x^2 + \frac{1}{4}x - 1 = 0$

Q. $\frac{1}{a}x^2 + \frac{1}{b}x + \frac{1}{c} = 0$

I. $2(m + 9)^2 + 3(m + 9) - 35 = 0$

R. $(n - 4)^2 - 7(n - 4) = -12$

2. Prove each statement, assuming $a, b, c \in \mathbf{R}$ and $a \neq 0$.
[Hint: Recall the solutions given by the quadratic formula.]

A. The sum of the solutions of the equation $ax^2 + bx + c = 0$ is $-b/a$.

B. The product of the solutions of the equation $ax^2 + bx + c = 0$ is c/a.

3. Use the properties provided in exercise 2 to determine a quadratic equation that has the given set of solutions.

A. $\{-9,\ 2\}$

C. $\{-6,\ -10\}$

B. $\{1 - \sqrt{5},\ 1 + \sqrt{5}\}$

D. $\{4 - \sqrt{7},\ 4 + \sqrt{7}\}$

Set 64: Quadratic Equations in One Variable

1. Solve each quadratic equation.

A. $x^2 - 3x - 40 = 0$

N. $x^2 + 16x + 63 = 0$

B. $x^2 - 14x + 40 = 0$

O. $x^2 - 64 = 0$

C. $v^2 + 10v = -25$

P. $w^2 - w = 42$

D. $m^2 = 3 - 2m$

Q. $n^2 = 12n - 32$

E. $x^2 + 5x - 1 = 0$

R. $x^2 - 7x + 4 = 0$

F. $8h^2 - 56 = 0$

S. $4k^2 = -10k$

G. $2x^2 + 6x - 3 = 0$

T. $3x^2 - 2x - 7 = 0$

H. $4x^2 - 11x + 5 = 0$

U. $5x^2 + 10x + 3 = 0$

I. $x^2 + x\sqrt{13} - 3 = 0$

V. $2x^2 - x\sqrt{89} + 5 = 0$

J. $7x^2 - x\sqrt{37} + 1 = 0$

W. $3x^2 + x\sqrt{33} - 4 = 0$

K. $4x - \dfrac{65}{x} - 7 = 0$

X. $3x + \dfrac{8}{x} = 14$

L. $1 + \dfrac{1}{x+4} + \dfrac{6}{x} = 0$

Y. $1 + \dfrac{4}{x+3} - \dfrac{12}{x} = 0$

M. $2 - \dfrac{1}{x-2} - \dfrac{2}{x} = 0$

Z. $2 + \dfrac{11}{x-5} + \dfrac{2}{x} = 0$

2. Solve each nonquadratic equation.

A. $|2x - 5| = |3x + 6|$

E. $|5x - 7| = |8 - 5x|$

B. $|x + 9| = |7x + 2|$

F. $|4x - 6| = |2x - 3|$

C. $|x^2 + 4x - 7| = |x^2 + 7x - 33|$

G. $|2x^2 - 13x - 1| = |2x^2 - 7x + 10|$

D. $|2x^2 + 2x + 1| = |x^2 + 9x + 9|$

H. $|3x^2 - x + 2| = |2x^2 + 9x - 23|$

Set 65: Word Problems: Quadratic Applications

1. Solve each word problem.

   A. The square of a negative number equals the difference of 6 and the number. Determine the number.

   B. Three less than two times the square of a positive number equals the sum of seven and eight times the number. Determine the number.

   C. The sum of 20 and the product of two consecutive even numbers is 100. Determine the numbers.

   D. The difference of the product of two consecutive odd numbers and 19 is 80. Determine the numbers.

   E. The length of a rectangle is three inches less than twice its width. The area of the rectangle is 104 square inches. Determine the dimensions of the rectangle.

   F. The base of a triangle is three-fourths of its height. The area of the triangle is 24 square feet. Determine the base and height of the triangle.

   G. A ladder leans against a house. The distance between the top of the ladder and the ground is two feet more than twice the distance between the bottom of the ladder and the house, while the length of the ladder is 13 feet. Determine the distance between the top of the ladder and the ground.

   H. An ant walks east from point A to point B, then south to point C, then directly back to point A. The distance between points A and C is four yards more than twice the distance between points A and B. The distance between points B and C is 16 yards. Determine the distance between points A and B.

   I. A frame of uniform width of 3 inches surrounds a rectangular painting. The length of the painting is 5 inches more than its width. The entire area of the painting and frame is 414 square inches. Determine the dimensions of the painting apart from the frame.

   J. The rectangular flag of a land of savages consists of a rectangle shaded red within a black border of uniform width. The dimensions of the red rectangle are 11 inches by 7 inches. The area of the flag is 165 square inches. Determine the width of the black border.

K. A square has area 81 square inches. Determine the length of the sides of the square.

L. An artist draws horizontal and vertical lines on a picture of a square to divide it into identical smaller squares. The length of the sides of the original square is 10 inches more than the length of the sides of a smaller square. The area of the original square is 144 square inches. Find the length of the sides of each of the smaller squares.

M. The base of a parallelogram is eight centimeters less than three times its height. The area of the parallelogram is 91 square centimeters. Determine the dimensions of the parallelogram.

N. A troop of 126 soldiers in the Prussian army lines up in even rows and columns in a rectangular formation. The number of rows of soldiers is nine more than twice the number of columns. Determine the number of rows of soldiers in the formation.

O. The sum of the areas of two circles is 136π square units. The radius of one circle is 4 units more than the radius of the other. Determine the radii of the circles.

P. The sum of the areas of two squares is 306 square meters. The length of the sides of one square is 6 meters more than the length of the sides of the other square. Find the length of the sides of the larger square.

Q. Bill drives 30 miles to work every morning at his dead-end job. One morning, a traffic jam forces Bill to drive 27 mph below his average speed. As a result, he arrives at work 15 minutes later than normal. Find the number of minutes it usually takes Bill to commute to work.

R. Carlos and Juan each must stack 70 boxes. Juan can stack boxes at a rate that is 2 boxes per hour slower than Carlos. As a result, Carlos can leave work 10 minutes earlier than Juan. Determine the length of time it takes Carlos to stack the boxes.

S. The product of two numbers is 13. The sum of the same numbers is 9. Identify the numbers.

T. The product of two numbers is 10. The sum of the same numbers is 8. Identify the numbers.

Set 66: Rational Exponents

1. Evaluate each numerical expression.

A. $36^{1/2}$

B. $-125^{1/3}$

C. $(-8)^{5/3}$

D. $-4^{7/2}$

E. $27^{-2/3}$

F. $3^{3/2} \cdot 3^{3/2}$

G. $11^{8/3} \cdot 11^{-2/3}$

H. $10^{7/3}/10^{1/3}$

I. $19^{1/2}/19^{-1/2}$

J. $-81^{1/2}$

K. $(-64)^{1/3}$

L. $27^{4/3}$

M. $64^{3/2}$

N. $8^{-4/3}$

O. $7^{1/2} \cdot 7^{3/2}$

P. $5^{15/4} \cdot 5^{-3/4}$

Q. $6^{5/2}/6^{1/2}$

R. $4^{4/3}/4^{-5/3}$

2. Simplify each expression and leave your answers with positive exponents. Assume that all variables represent positive real numbers.

A. $(7x^{3/2}y^{1/5})^2$

B. $(4m^{3/7})(-9m^{2/7})$

C. $(v^{-3/2})(8v^{2/5})$

D. $(12x^{3/5})/(4x^{2/3})$

E. $(6b^5)/(-15b^{11/4})$

F. $(2x^{-7/2}y^{1/5})^3$

G. $[(c^{1/3})(2c^{-1/4})]^{-5}$

H. $[m^{-3/8}/m^{2/3}]^4$

I. $\dfrac{(x^{2/3}y^{5/2})^4}{(x^{7/4}y^{-1/3})^{6/7}}$

J. $(3x^{1/4}y^{2/3})^3$

K. $(10x^{5/6}y^{-9/2})^2$

L. $(-8n^{5/3})(3n^{19/3})$

M. $(2w^{5/8})(9w^{6/5})$

N. $(-2x^4)/(10x^{1/3})$

O. $(-18k^{3/10})(3k^{1/4})$

P. $[(3x^{-5/2})(x^{2/5})]^3$

Q. $[n^{-1/6}/n^{3/4}]^{-2}$

R. $\dfrac{(x^5y^2)^{2/3}}{(x^{-2}y^7)^{1/5}}$

3. Use rational exponents to simplify each expression. Leave each answer as a radical.

A. $\sqrt[8]{x^2}$

B. $\sqrt[4]{9}$

C. $\sqrt{x} \cdot \sqrt[4]{x^3}$

D. $\sqrt[3]{m^4} / \sqrt[5]{m^2}$

E. $\sqrt[7]{9} \cdot \sqrt[4]{27}$

F. $\sqrt[15]{x^3}$

G. $\sqrt[6]{8}$

H. $\sqrt[3]{x^2} \cdot \sqrt[8]{x^5}$

I. $\sqrt[7]{n^8} / \sqrt[3]{n^3}$

J. $\sqrt[3]{4} / \sqrt[5]{8}$

4. Solve each equation.

A. $3x^{2/5} - 7 = 5$

B. $6 + n^{5/2}/8 = 10$

C. $-2x^{2/3} + 21 = 3$

D. $n^{3/4} + 4 = 31$

5. Evaluate each numerical expression.

A. $16^{-2^{-2}}$

B. $(-125)^{3^{-1}}$

Set 67: Factorizing by Subsitution

1. Factorize each expression completely.

A. $x^{2/5} + 5x^{1/5} + 6$

G. $x^{2/3} - 4x^{1/3} - 21$

B. $v^{-1} + 2v^{-1/2} - 24$

H. $w^{-11/2} - 10w^{-11/4} + 25$

C. $m^{5/2} - 9m^{1/2}$

I. $n^{3/2} - 16n^{1/2}$

D. $x^{4/3} + 27x^{1/3}$

J. $x^{9/5} - 8x^{6/5}$

E. $3(2x + 7)^2 - 2(2x + 7) - 16$

K. $6(5x - 3)^2 - 27(5x - 3) + 21$

F. $5(x - 9)^3 - 320$

L. $(10x + 3)^3 - 10x - 3$

2. Solve each equation.

A. $x + 10\sqrt{x} = 11$

G. $x^2 - x\sqrt{x} - 12x = 0$

B. $x^{5/2} - 49x^{3/2} = 0$

H. $3x^{9/2} = 48x^{5/2}$

C. $m^{-2} - 7m^{-1} + 10 = 0$

I. $n^{-6} + 8n^{-3} + 15 = 0$

D. $(x + 6)^2 - 6(x + 6) - 27 = 0$

J. $(4x - 7)^2 + 2(4x - 7) - 63 = 0$

E. $2x^4 - 6x^2 - 80 = 0$

K. $x^6 + 4x^3 = 45$

F. $v^{-8} - 13v^{-5} + 30v^{-2} = 0$

L. $w^{-6} - 81w^{-4} = 0$

3. Use the property of exercise 3C in set 31 to simplify the expression below:

$$(x - 1)^4 + 4(x - 1)^3 + 6(x - 1)^2 + 4(x - 1) + 1$$

Set 68: The Cartesian Plane

1. Construct a portion of the Cartesian plane that will contain the points below and then plot those points in the plane.

A. $A(5, 2)$ G. $G(4, -5)$

B. $B(-3, 0)$ H. $H(-6, -1)$

C. $C(-4, -6)$ I. $I(3, 8)$

D. $D(0, 1)$ J. $J(6, 0)$

E. $E(2, -2)$ K. $K(-5, 4)$

F. $F(-2, 7)$ L. $L(0, -4)$

2. Identify the quadrant of the Cartesian plane in which each point lies.

A. $M(-1, 9)$ E. $R(852, 119)$

B. $N(8, -5)$ F. $S(-\pi, 21e)$

C. $P(7, 7)$ G. $T(11/2, -\sqrt{5})$

D. $Q(-33, -2)$ H. $U(-0.034, -2.894)$

Set 69: Distance and Midpoint Formulas, Two Dimensions

1. Given the coordinates of points P and Q, determine:
   - the distance $d(P, Q)$ between the points and
   - the coordinates of the midpoint of the line segment PQ.

A. $P(4, 9)$, $Q(10, 1)$

I. $P(-5, 3)$, $Q(7, 8)$

B. $P(-11, -3)$, $Q(-14, -7)$

J. $P(0, -4)$, $Q(-15, 4)$

C. $P(-3, 8)$, $Q(-3, 9)$

K. $P(4, 5)$, $Q(-2, 5)$

D. $P(1, 7)$, $Q(-1, 10)$

L. $P(9, 4)$, $Q(4, 7)$

E. $P(-8, -1)$, $Q(-4, 5)$

M. $P(-15, -2)$, $Q(-12, -11)$

F. $P(-a, 0)$, $Q(0, b)$

N. $P(3a, 0)$, $Q(0, 7b)$

G. $P(3/2, -7/3)$, $Q(-5/2, -4/3)$

O. $P(-6/5, 9/5)$, $Q(2/5, 14/5)$

H. $P(-2/3, 5/2)$, $Q(1/7, 8/5)$

P. $P(7/3, 10/3)$, $Q(1/2, 7/2)$

2. Determine the area of the polygon described.

   A. the rectangle with vertices $(2, 3)$, $(11, 3)$, $(11, 7)$, and $(2, 7)$

   B. the triangle with vertices $(-12, -2)$, $(-5, -2)$, and $(-3, 6)$

   C. the parallelogram with vertices $(-4, -2)$, $(8, 3)$, $(11, 7)$, and $(-1, 2)$

3. Which lies closer to the point $P(3, -8)$: point $A(-1, -3)$ or point $B(9, -6)$?

4. Prove each statement.

   A. The midpoint of the hypotenuse of a right triangle is equidistant from all three vertices of the triangle.

   B. The three line segments that join the midpoints of the three sides of an isosceles triangle form an isosceles triangle themselves.

Set 70: Graphs of Equations in Two Variables

1. Solve each equation for the specified variable.

A. $2x - 5y = 10$ for y

D. $\frac{3}{2}x + 6y = -12$ for x

B. $\sqrt{y} - \sqrt{x} = 3$ for x

E. $4x + \sqrt[3]{5y} = 8$ for y

C. $x^2 + y^2 = 25$ for y

F. $y = (x-2)^2 - 9$ for x

2. Draw the graph of each equation.

A. $y = x - 3$

L. $x = -2y + 6$

B. $3x + 2y = 7$

M. $4x - \frac{1}{3}y = 1$

C. $x = -3$

N. $y = 2$

D. $x = (y-1)^2 - 8$

O. $y = -(x+2)^2 + 5$

E. $2x + \sqrt{y} = -4$

P. $3\sqrt{x} - y = -1$

F. $(x+3)^2 + (y-4)^2 = 9$

Q. $x^2 + (y+5)^2 = 16$

G. $y = (x+1)^3$

R. $x = 2(y-4)^3$

H. $|4y| = 12$

S. $|x| = 5$

I. $x = |7 - 2y|$

T. $y = |x+5|$

J. $|y| - |x| = 2$

U. $|x| + |y| = 7$

K. $9x^2 - y^2 = 0$

V. $y^2 - 36 = 0$

3. Identify the intercepts of the graph of each equation.

A. $6x + 5y = 15$

D. $3x - 2y = -24$

B. $y = x^3 + 8$

E. $(x-2)^2 - 4y^2 = 25$

C. $x^2 + y^2 = 64$

F. $y = x^2 - 7$

Set 71: Lines

1. Find the slope m of the line that passes through the points with the given coordinates.

A. $(-5, 2)$ and $(-3, 10)$

D. $(8, -1)$ and $(6, 11)$

B. $(0, 7)$ and $(3, 6)$

E. $(13, 4)$ and $(-5, 8)$

C. $(4, -2)$ and $(1, -2)$

F. $(-7, 3)$ and $(-7, 7)$

2. The line through the points $(1, 4)$ and $(x, 16)$ has slope -3. Find x.

3. The line through the points $(-4, -6)$ and $(3, y)$ has slope $1/2$. Find y.

4. The line through the points $(-\sqrt{2}, \sqrt{8})$ and $(\sqrt{18}, y)$ has slope 5. Find y.

5. The line through the points $(m, 12)$ and $(11, 7m)$ has slope m, where $m > 0$. Determine m.

6. Determine whether the slope of each line is positive, negative, zero, or undefined.

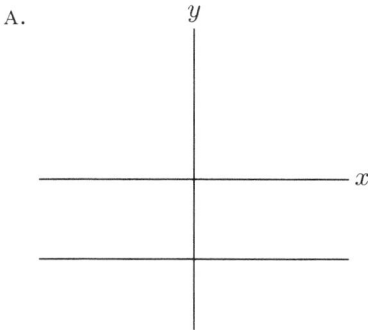

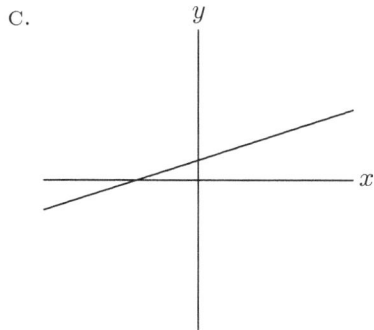

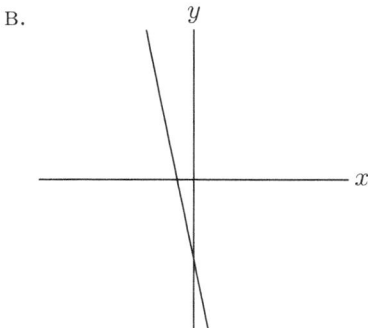

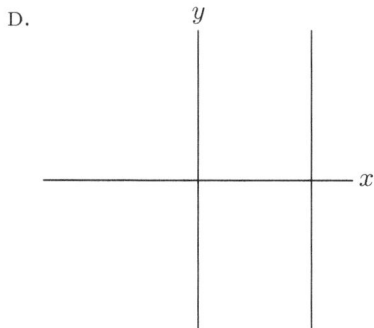

A.

C.

B.

D.

7. Given a point (a, b) on a line with slope m, find three other points on the line.

   A. $(a, b) = (0, 3)$, $m = 5/2$ D. $(a, b) = (-2, 7)$, $m = -2/3$

   B. $(a, b) = (-8, -8)$, $m = -7$ E. $(a, b) = (-1, -5)$, $m = 6$

   C. $(a, b) = (9, -4)$, $m = 0$ F. $(a, b) = (6, 1)$, slope is undefined

8. Find the coordinates of two points on the line with the given equation and then use those points to find the slope m of the line.

   A. $y = \frac{8}{3}x + 5$ E. $y = -\frac{1}{2}x - 7$

   B. $x = -5y - 8$ F. $x = \frac{7}{6}y + 11$

   C. $2x - 8y = 9$ G. $5x + 4y = -1$

   D. $y = -9$ H. $x = 12$

Set 72: Linear Equations in Two Variables

1. Determine the equation of the line with the given slope m and y-intercept b. Write the equation in both:
  - slope-intercept form $y = mx + b$;
  - standard form $Ax + By = C$, where $A, B, C \in \mathbf{Z}$ and $A \geq 0$.

A. $m = 4; \, b = -3$ D. $m = -3; \, b = 5$

B. $m = -5/3; \, b = 0$ E. $m = 2/7; \, b = 2$

C. $m = 9/2; \, b = 1/11$ F. $m = -8/5; \, b = -3/4$

2. Determine the equation of the line that contains the given point and has the given slope. Write the equation in both:
  - slope-intercept form $y = mx + b$ (if possible);
  - standard form $Ax + By = C$, where $A, B, C \in \mathbf{Z}$ and $A \geq 0$.

A. point: $(-3, 4); \, m = 5/6$ D. point: $(1, 5); \, m = 3/2$

B. point: $(0, -5); \, m = -8$ E. point: $(-9, -2); \, m = -1/4$

C. point: $(6, 3); \, m = 0$ F. point: $(1, 2);$ slope undefined

3. Determine the equation of the line on which the given points lie. Write the equation in both:
  - slope-intercept form $y = mx + b$ (if possible);
  - standard form $Ax + By = C$, where $A, B, C \in \mathbf{Z}$ and $A \geq 0$.

A. $(1, 2)$ and $(9, 6)$ D. $(-3, 0)$ and $(2, -2)$

B. $(-8, 5)$ and $(-10, 11)$ E. $(4, -3)$ and $(6, 14)$

C. $(-7, 2)$ and $(-7, -6)$ F. $(0, 7)$ and $(5, 7)$

4. Determine the intercepts of the graph of each equation and then draw that graph.

A. $x + 5y = 10$ E. $4x - 3y = -12$

B. $3x - 7y = 2$ F. $8x + y = -3$

C. $y = \frac{8}{3}x - 1$ G. $y = -\frac{3}{4}x + 2$

D. $x = 4$ H. $y = -3$

5. Derive the equation that relates the measure F of a temperature in degrees Fahrenheit to the measure C of the same temperature in degrees Celsius. Assume that $0°C = 32°F$ and $100°C = 212°F$ and leave the equation in the form $F = mC + b$.

6. Convert each linear equation to slope-intercept form (if necessary) and then determine the slope m and y-intercept b of the line with that equation.

A. $y = -5x + 1$

B. $y = x/3 - 7$

C. $y = 88x$

D. $7x - 2y + 3 = 0$

E. $\frac{1}{4}x + \frac{9}{5}y = -4$

F. $x/5 - y/2 = 10$

G. $y = 2x - 11$

H. $y = 6 - x$

I. $y = -23$

J. $x + 5y = 0$

K. $\frac{2}{3}x - \frac{1}{7}y = -2$

L. $3x/8 + 2y/11 = 1$

Set 73: Parallel and Perpendicular Lines

1. Given the slope m of a line, find:
- the slope $m_{\parallel}$ of any line parallel to the given line and
- the slope $m_{\perp}$ of any line perpendicular to the given line.

A. $m = -2/9$ D. $m = 1/4$

B. $m = 5$ E. $m = -8$

C. $m = 0$ F. m is undefined

2. Determine the equation of the line that satisfies the given conditions. Leave the equation in standard form.

A. passes through $(-4, -5)$;
is vertical

B. passes through $(0, 6)$;
is horizontal

C. passes through $(-2, 13)$;
runs parallel to the x-axis

D. passes through $(3, -5)$;
runs perpendicular to the y-axis

E. passes through $(8, -1)$;
is horizontal

F. passes through $(-7, 2)$;
is vertical

G. passes through $(9, 0)$;
runs parallel to the y-axis

H. passes through $(-3, -12)$;
runs perpendicular to the x-axis

3. Determine the equation of the line that satisfies the given conditions. Leave the equation in slope-intercept form.

A. passes through $(0, 6)$;
runs parallel to the line with
equation $y = 5x - 17$

B. passes through $(-11, 4)$;
runs perpendicular to the line
with equation $3x - 9y = -2$

C. passes through $(14, -3)$;
runs perpendicular to the line
with equation $x = 1$

D. passes through $(-1, -3)$;
runs parallel to the line through
$(7, 7)$ and $(-2, 5)$

E. passes through $(5, 1)$;
runs parallel to the line with
equation $8x - 5y = 20$

F. passes through $(-7, -8)$;
runs perpendicular to the line
with equation $y = -\frac{3}{2}x - 6$

G. passes through $(9, 15)$;
runs parallel to the line with
equation $y = -4$

H. passes through $(-10, 7)$;
runs perpendicular to the line
through $(-5, 0)$ and $(3, 1)$

4. Verify each statement.

A. Points $A(1,5)$, $B(-3,12)$, and $C(4,16)$ serve as the vertices of a right triangle.

B. Points $E(-3,-2)$, $F(2,0)$, $G(5,7)$, and $H(0,5)$ serve as the vertices of a parallelogram.

C. The diagonals of the quadrilateral with vertices $P(4,3)$, $Q(6,8)$, $R(8,8)$, and $S(11,4)$ lie perpendicular to each other.

5. Determine the equation of the perpendicular bisector of the line segment that joins the points $P(-1,2)$ and $Q(9,5)$. Leave the equation in standard form.

Set 74: Solving Systems of Linear Equations by Graphing

1. Identify the solution of each system of equations given their graphs. Assume that each solution is an ordered pair of integers.

A.

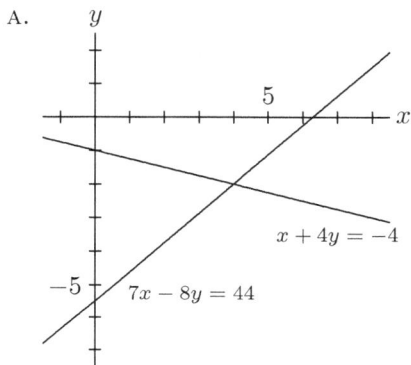

$x + 4y = -4$

$7x - 8y = 44$

C.

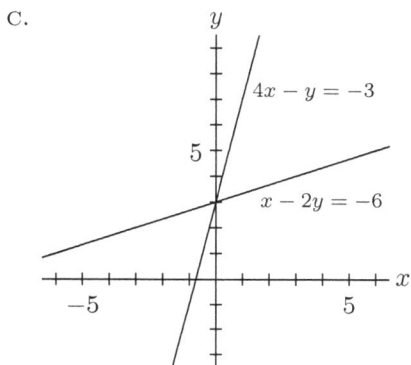

$4x - y = -3$

$x - 2y = -6$

B.

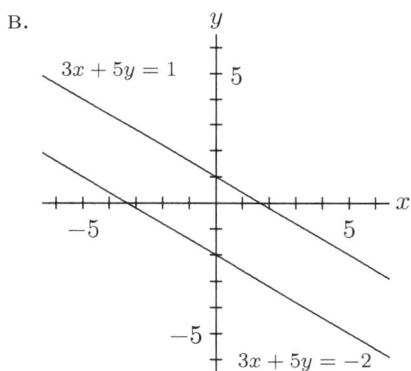

$3x + 5y = 1$

$3x + 5y = -2$

D.

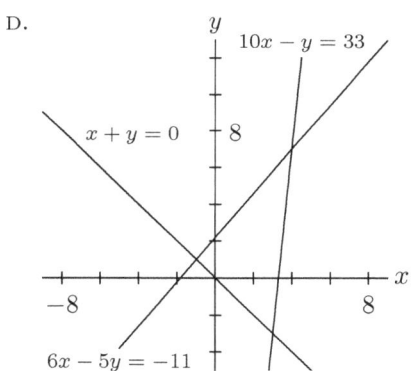

$10x - y = 33$

$x + y = 0$

$6x - 5y = -11$

2. For each system of equations:
  - draw the graphs of both equations in the system;
  - identify from the graphs the ordered pair that satisfies both equations;
  - verify that the ordered pair is the solution of the system.

Assume that each solution is an ordered pair of integers.

A. $\begin{cases} 3x + y = -1 \\ 5x - 3y = -25 \end{cases}$

B. $\begin{cases} 2x + 3y = 0 \\ 5x + 5y = 10 \end{cases}$

C. $\begin{cases} x = -6 \\ 4x - 3y = -21 \end{cases}$

D. $\begin{cases} 4x - 9y = 6 \\ x - 7y = 11 \end{cases}$

E. $\begin{cases} 6x - 5y = -4 \\ 3x - 8y = -13 \end{cases}$

F. $\begin{cases} 4x + 11y = 49 \\ y = 3 \end{cases}$

Set 75: Solving Systems of Linear Equations by Elimination

1. Determine constants to multiply to the sides of the equations in the given system in order to eliminate x from that system.

A. $\begin{cases} 2x - 6y = 36 \\ 7x + 4y = 1 \end{cases}$

B. $\begin{cases} 5x - 8y = -82 \\ 3x - 2y = -24 \end{cases}$

2. Determine constants to multiply to the sides of the equations in the given system in order to eliminate y from that system.

A. $\begin{cases} 2x - 6y = 36 \\ 7x + 4y = 1 \end{cases}$

B. $\begin{cases} 5x - 8y = -82 \\ 3x - 2y = -24 \end{cases}$

3. Solve each system of equations by elimination.

A. $\begin{cases} 3x + y = -1 \\ 5x - 3y = -25 \end{cases}$

J. $\begin{cases} 4x - 9y = -3 \\ x + 7y = -10 \end{cases}$

B. $\begin{cases} x + 8y = -7 \\ 2x + 5y = 8 \end{cases}$

K. $\begin{cases} 7x - y = -14 \\ 9x - 6y = -33 \end{cases}$

C. $\begin{cases} 2x - 6y = 36 \\ 7x + 4y = 1 \end{cases}$

L. $\begin{cases} 6x - 7y = -9 \\ 12x - 14y = -18 \end{cases}$

D. $\begin{cases} 5x + 2y = 3 \\ 15x + 3y = 8 \end{cases}$

M. $\begin{cases} 5x - 8y = -82 \\ 3x - 2y = -24 \end{cases}$

E. $\begin{cases} 6x - 9y = 14 \\ 4x - 6y = 10 \end{cases}$

N. $\begin{cases} 2x + 3y = 0 \\ 5x + 5y = 10 \end{cases}$

F. $\begin{cases} 9x - 10y = -10 \\ 6x + 2y = -11 \end{cases}$

O. $\begin{cases} 3x + 11y = 14 \\ 8x - 33y = 9 \end{cases}$

G. $\begin{cases} y = 7x - 9 \\ y = 2x + 1 \end{cases}$

P. $\begin{cases} y = -3x - 14 \\ y = 3x + 4 \end{cases}$

H. $\begin{cases} 6x - 7y = 22 \\ x - \frac{5}{2}y = -7 \end{cases}$

Q. $\begin{cases} \frac{2}{7}x - \frac{1}{3}y = 14 \\ \frac{3}{4}x - \frac{9}{2}y = 15 \end{cases}$

I. $\begin{cases} \frac{1}{2}x + \frac{3}{5}y = 6 \\ \frac{2}{3}x + \frac{7}{4}y = 27 \end{cases}$

R. $\begin{cases} 5x + 3y = -60 \\ \frac{1}{3}x - y = 8 \end{cases}$

Set 76: Solving Systems of Linear Equations by Substitution

1. Solve each system of equations by substitution.

A. $\begin{cases} 3x + y = 23 \\ 5x - 3y = 29 \end{cases}$
    G. $\begin{cases} 4x - 9y = 3 \\ x + 7y = 10 \end{cases}$

B. $\begin{cases} x + 8y = -43 \\ 2x + 5y = -20 \end{cases}$
    H. $\begin{cases} 7x - y = -21 \\ 9x - 6y = 6 \end{cases}$

C. $\begin{cases} 2x - 6y = 16 \\ 7x + 4y = -19 \end{cases}$
    I. $\begin{cases} 5x - 8y = 42 \\ 3x - 2y = 14 \end{cases}$

D. $\begin{cases} y = 7x + 17 \\ y = -2x - 1 \end{cases}$
    J. $\begin{cases} y = -3x + 8 \\ y = 3x - 28 \end{cases}$

E. $\begin{cases} -12x + 2y = 8 \\ 6x - y = -4 \end{cases}$
    K. $\begin{cases} 3x - 21y = -5 \\ x - 7y = -2 \end{cases}$

F. $\begin{cases} 6x - 7y = 26 \\ x - \frac{5}{2}y = 15 \end{cases}$
    L. $\begin{cases} 5x + 3y = 3 \\ \frac{1}{3}x - y = -19 \end{cases}$

2. Solve each system of nonlinear equations.

A. $\begin{cases} y = x^2 + 2x - 2 \\ y = x \end{cases}$
    F. $\begin{cases} 4x = y^2 + 6y + 17 \\ 2x = 2y^2 + 3y + 7 \end{cases}$

B. $\begin{cases} (x-1)^2 = 25 - (y+2)^2 \\ (x-1)^2 = 3 - y \end{cases}$
    G. $\begin{cases} (x-1)^2 + (y-2)^2 = 25 \\ (x+5)^2 + (y+4)^2 = 13 \end{cases}$

C. $\begin{cases} x^2 + y^2 = 25 \\ x + y = 1 \end{cases}$
    H. $\begin{cases} 9x^2 + 4y^2 = 36 \\ x - y = 3 \end{cases}$

D. $\begin{cases} x^2 + y = 10 \\ 3x - y = 0 \end{cases}$
    I. $\begin{cases} 2x - y^2 = -11 \\ 4x + 7y = 17 \end{cases}$

E. $\begin{cases} |x| - x + y = 18 \\ x + |y| + y = 11 \end{cases}$
    J. $\begin{cases} |x| - x - y = 5 \\ x + |y| - y = 5 \end{cases}$

3. Solve each word problem. Use the method of elimination or the method of substitution.

A. The sum of two numbers is 23. The difference of the numbers is 8. Determine the numbers.

B. The sum of the digits of a two-digit number is 10. The number yielded by reversing the two digits is 36 more than the original number. Determine the original number.

C. Oscar and Philip took an examination. The sum of their scores was 119 and the difference of their scores was 37. If Oscar's score was higher than Philip's score, then what were the scores?

D. Mona has 37 coins, each of which is either a dime (worth ten cents) or a quarter (worth twenty-five cents). The value of the coins is $7.15. Determine the number of coins of each kind in her collection.

E. Nick takes yellow pills and black pills. The total weight of three yellow pills and nine black pills is 84 grams. The total weight of seven yellow pills and five black pills is 68 grams. Determine the weight of each kind of pill.

F. To travel to Billings by the fast train, three adults and four children must pay $101 in total, whereas two adults and five children must pay $93 in total. Find the price of a ticket for an adult and the price of a ticket for a child.

G. Santiago and his wife Rosa work at the same camp, but for different hourly wages. At the end of each day they combine their earnings. On Monday, Rosa works seven hours and Santiago works five hours, and together they earn $66.45. On Tuesday, Rosa works nine hours and Santiago works four hours, and together they earn $64.55. What is the hourly wage of each laborer?

H. Two angles are complementary. The measure of one angle is $8°$ more than the measure of the other angle. Determine the measures of the angles.

I. Julius divides $800 into two bank accounts. He earns simple interest at a rate of 4% in one account and simple interest at a rate of 5% in the other account. After one year, the total amount of interest he has accrued in the accounts is $34. How much money did Julius deposit originally in each account?

J. The areas of the top, front, and left faces of a prism are 9 in², 15 in², and 60 in², respectively. Determine the volume of the prism.

4. Suppose $x - y = 3$ and $x^2 - y^2 = 15$. Determine $x + y$.

5. Suppose $x + y = 2$ and $x^3 + y^3 = 14$. Determine $x^2 - xy + y^2$.

Set 77: Linear Inequalities in Two Variables

1. Draw the graph of each inequality.

A. $y \geq x + 2$　　　　　　　　　　E. $y \geq -2x - 5$

B. $3x + 4y < 10$　　　　　　　　　F. $x - 3y < 6$

C. $5x - 2y > -6$　　　　　　　　　G. $7x + 3y < -11$

D. $x \geq -4$　　　　　　　　　　　H. $y \leq 1$

2. Draw the graph of the set of solutions of each system of linear inequalities.

A. $\begin{cases} y \leq 3x - 1 \\ y > -x/2 + 5 \end{cases}$　　　　　D. $\begin{cases} y \leq x + 4 \\ y \geq 2x - 3 \end{cases}$

B. $\begin{cases} 7x + 3y < 0 \\ 2x - 4y < 6 \end{cases}$　　　　　　E. $\begin{cases} x - 5y > -2 \\ 9x + 3y > 4 \end{cases}$

C. $\begin{cases} 5x - 2y > -1 \\ y \geq -3 \end{cases}$　　　　　　F. $\begin{cases} x < 2 \\ 4x - 8y \leq -7 \end{cases}$

3. Draw the graph of each nonlinear inequality.

A. $|y| < 3$　　　　　　　　　　　C. $|5x| \geq 10$

B. $y \geq |x - 1|$　　　　　　　　　D. $|x| + |y| < 4$

Set 78: Linear Equations in Three Variables

1. Solve each system of equations.

A.
$$\begin{cases} x - 3y + 2z = -4 \\ 2x - 4y + 3z = -5 \\ 5x + 2y + 4z = -7 \end{cases}$$

F.
$$\begin{cases} 3x - 6y + 3z = 0 \\ 4x + y - z = 1 \\ 2x - 7y + 5z = 9 \end{cases}$$

B.
$$\begin{cases} 4x - 7z = 5 \\ 4x + 6y + 9z = 5 \\ 3x - 3y - 8z = -12 \end{cases}$$

G.
$$\begin{cases} 7x + 2y - z = 7 \\ 3x - y + z = -8 \\ 5x - 2y = 2 \end{cases}$$

C.
$$\begin{cases} 6x + 5z = 10 \\ -y - z = -6 \\ 9x - 4y = 5 \end{cases}$$

H.
$$\begin{cases} 5x + 2y + 4z = -10 \\ 2x + 5z = -2 \\ 4y - 8z = 8 \end{cases}$$

D.
$$\begin{cases} -8x + 4y - 2z = -14 \\ 12x - 6y + 3z = 21 \\ 4x - 2y + z = 7 \end{cases}$$

I.
$$\begin{cases} 4x + 3y - 8z = 11 \\ x - 3y - 7z = 2 \\ 5x - 15y - 35z = -9 \end{cases}$$

E.
$$\begin{cases} 3x + y + 4z = 4 \\ -6x - 2y - 8z = -8 \\ 12x + 4y + 16z = 11 \end{cases}$$

J.
$$\begin{cases} 3x + 8y - 4z = -28 \\ -2x - 5y + z = 9 \\ 2x + 5y - z = -9 \end{cases}$$

2. Solve each word problem.

A. The hundreds digit of a three-digit number is twice the ones digit, the tens digit is two more than the hundreds digit, and the reverse of the number is 297 less than the number itself. Determine the number.

B. The sum of three numbers is 56, the difference of the largest and the smallest numbers is 6 more than the middle number, and the smallest number is 11 less than the middle number. Find the three numbers.

C. The measure of angle A in $\triangle ABC$ is five times the measure of angle B. The measure of angle C is 19° more than the measure of angle B. Find the measures of the three angles.

D. Kenneth mixes saline solutions with 10%, 20%, and 40% salt to obtain 50 gallons of a solution that is 29% salt. The volume of the 10% solution is three times the volume of the 20% solution. Determine the number of gallons of each solution that he mixed.

E. The sums of the numbers in the three subsets of size 2 of a given set of three numbers are 1, 7, and 16. Determine the given set.

Set 79: Symmetry

1. Prove each statement.

A. The graph of $y = -x^4$ is symmetric with respect to the y-axis.

B. The graph of $y = x^2 - 7$ is symmetric with respect to the y-axis.

C. The graph of $x = y^2 + 3$ is symmetric with respect to the x-axis.

D. The graph of $x = 10 - |y|$ is symmetric with respect to the x-axis.

E. The graph of $xy = 8$ is symmetric with respect to the origin.

F. The graph of $x^2 + 4y^2 = 4$ is symmetric with respect to the origin.

2. Provide a counterexample to disprove each statement.

A. The graph of $y = x^2 - 2x - 3$ is symmetric with respect to the y-axis.

B. The graph of $x = 5y^3 + 2$ is symmetric with respect to the x-axis.

C. The graph of $2x + 3y = 6$ is symmetric with respect to the origin.

3. Draw the graph of each equation and determine whether that graph has any symmetries with respect to the x-axis, y-axis, or origin.

A. $y = 6x$

E. $x = 2 - y^4/4$

B. $y = |x|/3 + 5$

F. $x^2 + y^2 = 4$

C. $y^2 = x - 7$

G. $4x + y = -8$

D. $y = (x - 1)^2$

H. $x^2 + y = 3$

PART TWO

Set 80: Functions

1. Suppose $f : \mathbf{R} \to \mathbf{R}$ is defined by $f(x) = 2x - 7$. Evaluate each expression.

  A. $f(4)$

  B. $f(0)$

  C. $f(e)$

  D. $f(n + 5)$

  E. $f(-3)$

  F. $f(\frac{3}{4})$

  G. $f(5m)$

  H. $f(x^2 - x + 1)$

2. Suppose $g : \mathbf{R} \to \mathbf{R}$ is defined by $g(x) = x^2 + 5x + 2$. Evaluate each expression.

  A. $g(-5)$

  B. $g(-\frac{1}{2})$

  C. $g(2y)$

  D. $g(\sqrt{v - 8})$

  E. $g(10)$

  F. $g(6z - 1)$

  G. $g(w^3 - 2w^2 + 4w - 5)$

  H. $g(\pi)$

3. Suppose $h : \mathbf{R} \to \mathbf{R}$ is defined piecewise by $h(x) = \begin{cases} 7x - 2 & \text{if } x < 3 \\ \sqrt{x + 6} & \text{if } x \geq 3 \end{cases}$. Evaluate each expression.

  A. $h(1)$

  B. $h(3)$

  C. $h(19)$

  D. $h(-5)$

4. Suppose $f : \mathbf{R} \to \mathbf{R}$ is defined piecewise by $f(x) = \begin{cases} x^2 + x & \text{if } x \leq 0 \\ |3x - 8| & \text{if } x > 0 \end{cases}$. Evaluate each expression.

  A. $f(4)$

  B. $f(-7)$

  C. $f(9)$

  D. $f(0)$

5. Suppose $g : \mathbf{R} \to \mathbf{R}$ is defined piecewise by $g(x) = \begin{cases} 2x^2 & \text{if } x \leq -5 \\ 7 & \text{if } -5 < x \leq 2 \\ 14 - x & \text{if } x > 2 \end{cases}$. Evaluate each expression.

  A. $g(-4)$

  B. $g(2)$

  C. $g(-11)$

  D. $g(40)$

  E. $g(-5)$

  F. $g(0)$

6. Suppose $h : (1, \infty) \to \mathbf{R}$ is defined by $h(x) = \begin{cases} 4x + 9 & \text{if } 1 < x < 8 \\ 7\sqrt{x} & \text{if } 8 \le x \le 20 \\ x & \text{if } x > 20 \end{cases}$.
Evaluate each expression.

A. $h(35)$ D. $h(9)$

B. $h(-2)$ E. $h(4)$

C. $h(17)$ F. $h(1)$

7. For each definition of f, determine the difference quotient $\dfrac{f(a + h) - f(a)}{h}$.
Be sure to simplify your answers completely.

A. $f(x) = 8x - 23$ D. $f(x) = 5x^2$

B. $f(x) = x^2 + 4x + 11$ E. $f(x) = x^3$

C. $f(x) = 4/(x - 1)$ F. $f(x) = \sqrt{3x}$

8. Suppose $f(x) = \dfrac{px + q}{rx - p}$ where $p^2 + qr \neq 0$. Determine $f\left(\dfrac{px + q}{rx - p}\right)$.

Set 81: Relations

1. For each relation, determine:
   - its domain;
   - its range;
   - whether or not it is a function.

   A. $\{(-5, 2), (-1, -3), (0, 8), (2, 7), (3, 10)\}$

   B. $\{(2, 6), (4, 7), (1, 9), (7, 11), (0, 13), (4, 16), (5, 8), (12, 5)\}$

   C. $\{(6, -1), (6, 2), (6, 10)\}$

   D. $\{(-9, 5), (-4, 5), (3, 5), (5, 5), (11, 5)\}$

Set 82: Domains and Range of Functions

1. Determine the domain $\mathbb{D}$ of the function with the given definition.

A. $f(x) = x + 7$

B. $g(x) = 6/(x - 3)$

C. $h(x) = \sqrt{10x + 3}$

D. $f(x) = \sqrt[5]{-4 - x}$

E. $g(x) = (3x + 1)/[(2x - 8)(4x + 7)]$

F. $h(x) = (9x - 2)/\sqrt{5x - 10}$

G. $f(x) = \sqrt{x^2 + 3x - 18}$

H. $g(x) = \sqrt{(x^2 - 9x + 14)/(x - 7)}$

I. $h(x) = \sqrt{(x + 5)(3x^2 - 27)}$

J. $h(x) = 14x/[(x + 9)(x - 5)]$

K. $f(x) = x^3 - 5x^2 - 8x + 11$

L. $g(x) = \sqrt[4]{2 - x}$

M. $h(x) = (x^2 - 5x + 4)/(x^2 - 2x - 8)$

N. $f(x) = \sqrt[3]{10x + 7}$

O. $g(x) = 11/\sqrt{4x + 6}$

P. $h(x) = 7/\sqrt{2x^2 + 11x + 15}$

Q. $f(x) = \sqrt{x - 17} - \sqrt{5x + 3}$

R. $g(x) = \sqrt{(2x^2 - 8)/(x^2 - x - 2)}$

2. Determine the range $\mathbb{R}$ of the function with the given definition.

A. $f(x) = x + 7$

B. $g(x) = 6/(x - 3)$

C. $h(x) = (5x + 1)/(2x - 4)$

D. $f(x) = \sqrt[5]{-4 - 9x}$

E. $g(x) = \sqrt{10x + 3}$

F. $h(x) = 12 - 2x^2$

G. $f(x) = 5x + 8$
where $f : [-6, -1] \rightarrow \mathbf{R}$

H. $g(x) = x^3 - 11$

I. $h(x) = 13x/(x + 2)$

J. $f(x) = (7x - 35)/(2x - 10)$

K. $g(x) = \sqrt[3]{8x + 7}$

L. $h(x) = \sqrt[4]{2 - x}$

M. $f(x) = x^2 + 2x - 5$

N. $g(x) = x^2 + 10$
where $g : (-3, 4] \rightarrow \mathbf{R}$

Set 83: Formulations of Functions

1. For each equation that relates x and y, determine the function f for which $y = f(x)$.

A. $5x - 3y = 15$

D. $8x^2 + 4y = 5$

B. $4xy = 1$

E. $8x/y = 14$

C. $x = y/(9 + y)$

F. $2xy - 11y = -7$

2. Define each function described.

A. V, where $V(r)$ represents the volume of a sphere with radius r

B. C, where $C(r)$ represents the circumference of a circle with radius r

C. A, where $A(b)$ represents the area of a triangle with base b and height 8

D. A, where $A(h)$ represents the area of a triangle with base 5 and height h

E. d, where $d(t)$ represents the distance traveled after t hours by an object moving at a speed of 60 miles per hour

F. I, where $I(t)$ represents the total amount of simple interest earned after t years on a deposit of \$450 at an annual rate of 2.5%

G. l, where $l(w)$ represents the length of a rectangle with an area of 100 and a width of w

H. S, where $S(V)$ represents the surface area of a cube with volume V

I. w, where $w(l)$ represents the width of a box with a surface area 216, height 6, and length l

J. A, where $A(C)$ represents the area of a circle with circumference C

3. Suppose a circle circumscribes a rectangle whose length is thrice its width. Define a function A where $A(w)$ represents the area of the circle if the width of the rectangle is w.

Set 84: Graphs of Functions

1. Determine whether each curve is the graph of a function.

A.

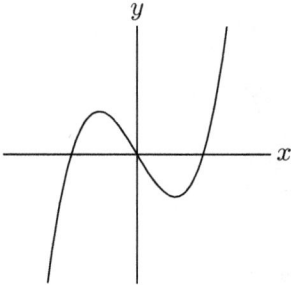

E.

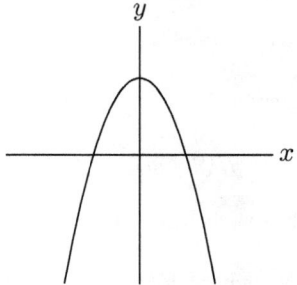

B.

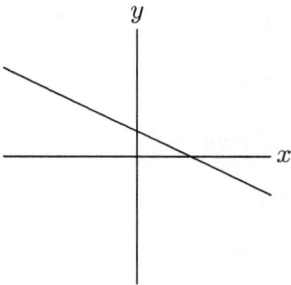

F.

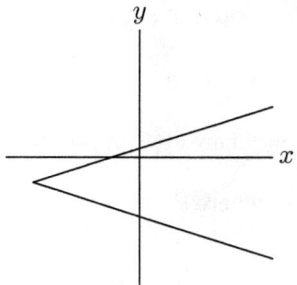

C.

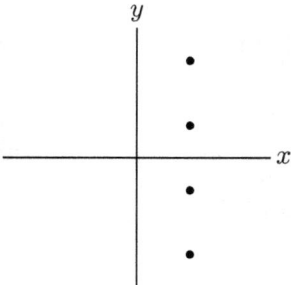

G.

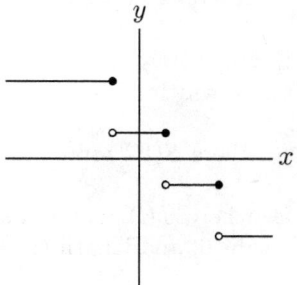

D.

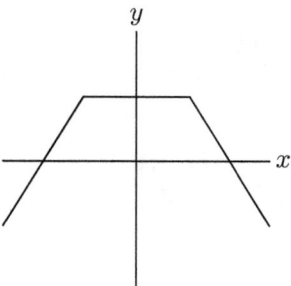

H.

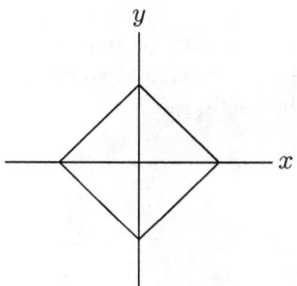

2. Draw the graph of the function with the given definition.

A. $f(x) = 8x - 3$

I. $g(x) = 1 - 5x/8$

B. $g(x) = 7$

J. $h(x) = -e$

C. $h(x) = x^2 - 2x + 1$

K. $f(x) = 2x^2 + 3x - 9$

D. $f(x) = x^3/5 + x + 10$

L. $g(x) = -4x^4 + 7$

E. $g(x) = |x + 7|$

M. $h(x) = \sqrt{3x + 12}$

F. $h(x) = 4\sqrt{6 - 2x}$

N. $f(x) = |4x - 1|/5$

G. $f(x) = \begin{cases} x/2 + 3 & \text{if } x \le 2 \\ -x^2 + 4x & \text{if } x > 2 \end{cases}$

O. $g(x) = \begin{cases} -3x & \text{if } x < 1 \\ \sqrt{x} - 2 & \text{if } x \ge 1 \end{cases}$

H. $h(x) = \begin{cases} -1 & \text{if } x < 0 \\ 0 & \text{if } x = 0 \\ 1 & \text{if } x > 0 \end{cases}$

P. $f(x) = \begin{cases} -x - 3 & \text{if } x \le -2 \\ -4 & \text{if } -2 < x \le 2 \\ 5x - 19 & \text{if } x > 2 \end{cases}$

Q. $g(x) = \lfloor x \rfloor$, where $\lfloor x \rfloor$ is the greatest integer less than or equal to x.

COMMENTS
- The function h of exercise 2H is called the *signum function*.
- The function g of exercise 2Q is called the *greatest integer function*.

Set 85: Characteristics of Functions

1. Given the graph of a function, identify:
   - the turning points in the graph,
   - the intervals on which the function is increasing and decreasing, and
   - the extrema of the function and the points in its domain where it attains those values.

A.

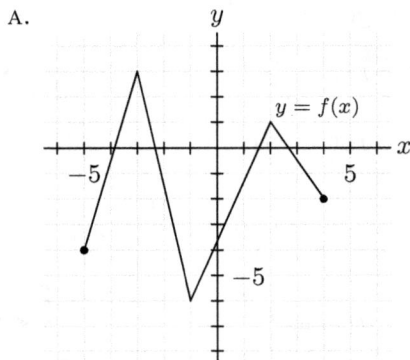

$y = f(x)$

C.

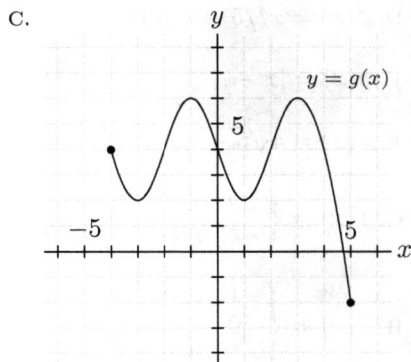

$y = g(x)$

B.

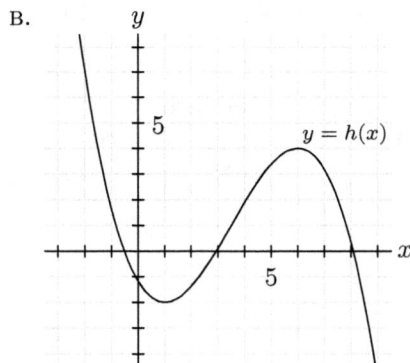

$y = h(x)$

D.

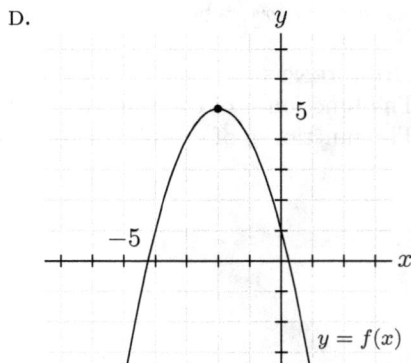

$y = f(x)$

2. Determine the rate of change of the given function on the given interval.

A. $f(x) = 6x + 1$; $[5, 9]$

B. $h(x) = 3x^2 - 7$; $[-8, -2]$

C. $g(x) = |x| - 11$; $[-1, 6]$

D. $f(x) = x^5 - 4$; $[-2, -1]$

E. $h(x) = x^{2/3}$; $[-8, 27]$

F. $g(x) = -9 - x$; $[-4, 3]$

G. $f(x) = 2x^2 + 8x + 5$; $[0, 7]$

H. $h(x) = -12 - 3\sqrt{x}$; $[4, 81]$

I. $g(x) = 4x^3 + 1$; $[2, 3]$

J. $f(x) = -5$; $[12, 17]$

Set 86: Transformations of Graphs of Functions

1. Suppose points $(-6, 10)$, $(0, -1)$, and $(3, 0)$ lie on the graph of function f. Translate these points to the graph of the function g as defined.

A. $g(x) = f(x - 7)$

B. $g(x) = 3 + f(x)$

C. $g(x) = -5 \cdot f(x)$

D. $g(x) = f(x/6)$

E. $g(x) = f(x + 4) - 9$

F. $g(x) = f(2x)/5$

G. $g(x) = 4 \cdot f(-4(x + 1)) - 11$

H. $g(x) = f(x) - 2$

I. $g(x) = f(x + 10)$

J. $g(x) = f(-3x)$

K. $g(x) = \frac{1}{2} \cdot f(x)$

L. $g(x) = 7 + f(x - 8)$

M. $g(x) = 9 \cdot f(x/4)$

N. $g(x) = -f(6x - 10) + 4$

2. For each equation:
- determine the points on its graph that correspond to points shown on the graph of $y = |x|$, and
- draw its graph.

A. $y = |x + 5|$

B. $y = \frac{1}{2} \cdot |x| + 4$

C. $y = -2 \cdot |-3x - 9| - 1$

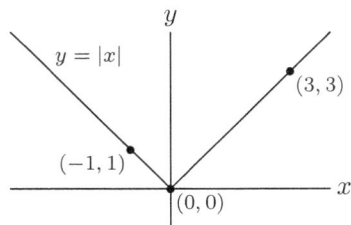

3. For each equation:
- determine the points on its graph that correspond to points shown on the graph of $y = x^3$, and
- draw its graph.

A. $y = -2x^3$

B. $y = (x/3)^3 - 7$

C. $y = -(4(x - 5))^3 + 2$

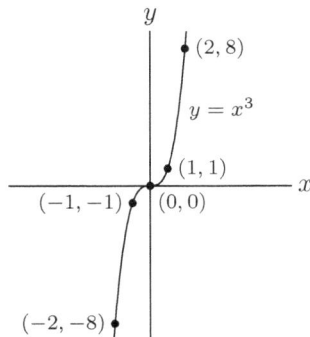

4. For each function:
   - determine the points on its graph that correspond to points shown on the graph of $y = f(x)$, and
   - draw its graph.

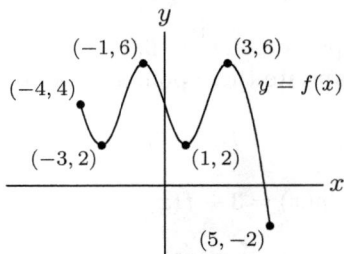

A. $g(x) = f(x) - 1$

B. $g(x) = f(2x - 8)$

C. $g(x) = 4f(7 - x) + 3$

5. For each function:
   - determine the points on its graph that correspond to points shown on the graph of $y = f(x)$, and
   - draw its graph.

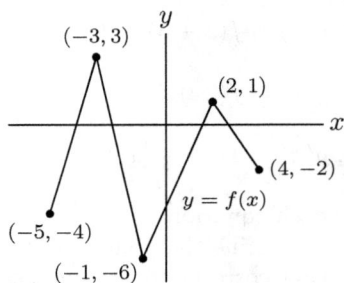

A. $h(x) = f(-4x)$

B. $h(x) = 3f(x + 9)$

C. $h(x) = \frac{1}{10}f(2(x - 1)) - 8$

Set 87: Translations of Graphs of Equations

1. For each equation:
  - find the points on its graph that correspond to points shown on the graph of $2x = -|y|$, and
  - draw its graph.

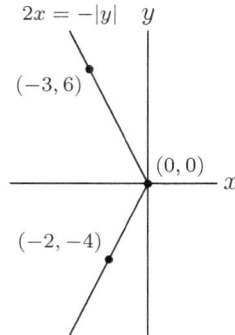

$$2x = -|y| \quad y$$

$(-3, 6)$

$(0, 0)$

x

$(-2, -4)$

A. $2x = -|y - 5|$

B. $2(x + 3) = -|y|$

C. $2x - 14 = -|y + 2|$

2. For each equation:
  - find the points on its graph that correspond to points shown on the graph of $x^2 + y^2 = 9$, and
  - draw its graph.

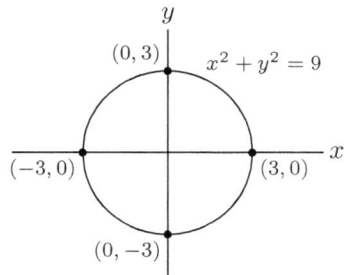

y

$(0, 3)$

$x^2 + y^2 = 9$

$(-3, 0)$

x

$(3, 0)$

$(0, -3)$

A. $x^2 + (y + 4)^2 = 9$

B. $(x - 1)^2 + (y - 9)^2 = 9$

C. $x^2 + 4x + y^2 - 10y + 20 = 0$

3. For each equation:
  - find the points on its graph that correspond to points shown on the graph of $x^2/4 + y^2/25 = 1$, and
  - draw its graph.

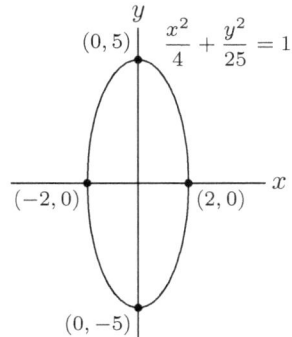

y

$(0, 5)$

$$\frac{x^2}{4} + \frac{y^2}{25} = 1$$

$(-2, 0)$

x

$(2, 0)$

$(0, -5)$

A. $(x - 3)^2/4 + y^2/25 = 1$

B. $x^2/4 + (y - 10)^2/25 = 1$

C. $(x + 2)^2/4 + (y + 8)^2/25 = 1$

Set 88: Even and Odd Functions

1. Determine whether each function is even, odd, or neither even nor odd, and provide a proof or counterexample to justify your conclusion.

A. $f(x) = x^2$

B. $g(x) = 3x^3 + 9x$

C. $h(x) = x^2 - 3x + 8$

D. $f(x) = 4$

E. $g(x) = |9x|$

F. $h(x) = |x| - x$

G. $f(x) = -\dfrac{7}{x}$

H. $h(x) = 4x^5$

I. $f(x) = 2x^4 - 7x^2 - 5$

J. $g(x) = 5x + 11$

K. $h(x) = \sqrt{x}$

L. $f(x) = |x - 1|$

M. $g(x) = 11|x| - 2$

N. $h(x) = \dfrac{8}{(x-3)(x+3)}$

O. $g(x) = \begin{cases} 4 & \text{if } x < 0 \\ 0 & \text{if } x = 0 \\ 4 & \text{if } x > 0 \end{cases}$

P. $f(x) = \begin{cases} \dfrac{x^2 + 6x}{2x + 12} & \text{if } x \neq -6 \\ -3 & \text{if } x = -6 \end{cases}$

2. Does a function exist that is both even and odd? If so, identify one.

3. Evaluate $f(5)$ for each function f where $f(-5) = 3$ and a, b, c, d, and e are real constants.

A. $f(x) = ax^9 + bx^7 + cx^5 + dx^3 + ex + 8$

B. $f(x) = ax^8 + bx^6 + cx^4 + dx^2 - 2x + 1$

Set 89: One-to-One Functions

1. Determine whether each curve is the graph of a one-to-one function.

A.

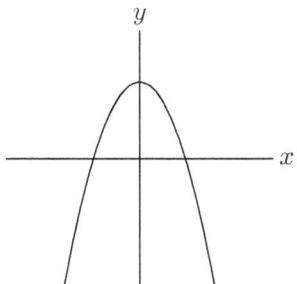

E.

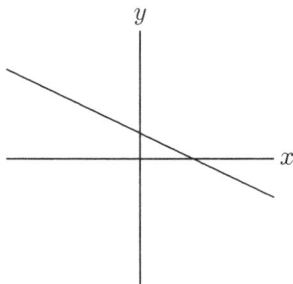

B.

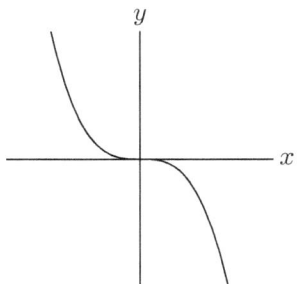

F.

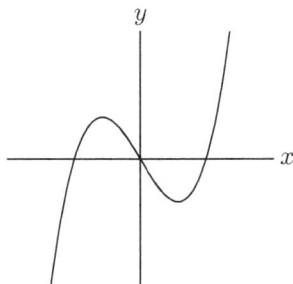

C.

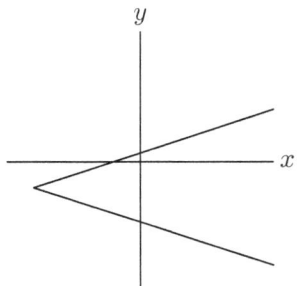

G.

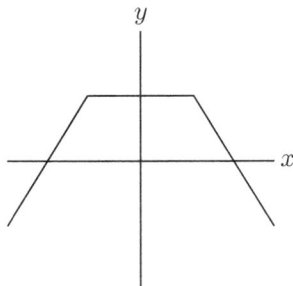

D.

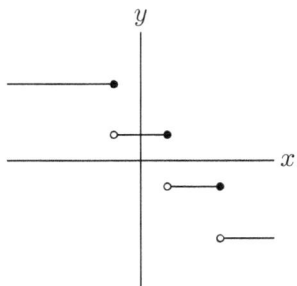

H.

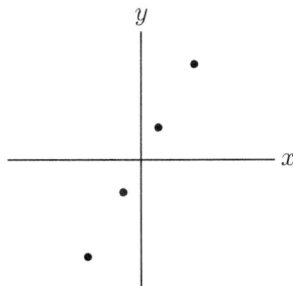

2. Determine whether each function is one-to-one or not one-to-one. Provide a proof or counterexample to justify your conclusion.

A. $f(x) = 9x^2$

E. $g(x) = -8x^3 + 1$

B. $g(x) = 10 - 3x$

F. $h(x) = 6$

C. $h(x) = x^3 - 4x$

G. $f(x) = \sqrt{x + 3}$

D. $f(x) = -(\sqrt{x})^4$

H. $g(x) = 5|x|$

3. Does a function exist that is both even and one-to-one? If so, identify one.

Set 90: Combinations of Functions

1. Suppose f and g are functions where:

$$f(-4) = 6 \qquad f(0) = 1 \qquad f(1) = -2 \qquad f(3) = 0 \qquad f(6) = 5$$
$$g(-4) = 7 \qquad g(-2) = 6 \qquad g(0) = 3 \qquad g(3) = 1 \qquad g(5) = -4$$

Evaluate each expression.

A. $(f + g)(3)$

B. $(fg)(0)$

C. $(g - f)(-4)$

D. $(g/f)(3)$

E. $(f \circ g)(-2)$

F. $(g \circ g)(5)$

G. $(f \circ g \circ f)(6)$

H. $(f - g)(0)$

I. $(f/g)(-4)$

J. $(g + f)(3)$

K. $(gf)(-4)$

L. $(f \circ f)(3)$

M. $(g \circ f)(1)$

N. $(g \circ f \circ g)(0)$

2. Given the definitions of functions f and g:
- determine formulas that define $f + g$, $f - g$, fg, f/g, $f \circ g$, and $g \circ f$;
- determine the domains of $f + g$, $f - g$, fg, f/g, $f \circ g$, and $g \circ f$;
- evaluate $(f + g)(-3)$, $(f - g)(6)$, $(fg)(2)$, and $(f/g)(5)$;
- evaluate $(f \circ g)(-2)$ and $(g \circ f)(3)$ in two ways:
  - using the formulas determined for $f \circ g$ and $g \circ f$;
  - using the definition of composition directly.

A. $f(x) = 4x + 7$;
   $g(x) = 8 - x$

B. $f(x) = x - 5$;
   $g(x) = 1/(2x^2 - 7x - 15)$

C. $f(x) = |7x + 1|$;
   $g(x) = -6$

D. $f(x) = (3x - 5)/(x + 2)$;
   $g(x) = x/(x + 2)$

E. $f(x) = x^2 + x - 6$;
   $g(x) = x + 3$

F. $f(x) = \sqrt{x + 9}$;
   $g(x) = x^2 - 4x - 5$

G. $f(x) = 6/x$;
   $g(x) = 3/(2x)$

H. $f(x) = (x - 7)/4$;
   $g(x) = 4x + 7$

3. Determine three different functions f for which $(f \circ f)(x) = x$.

4. Determine all linear functions f for which $(f \circ f)(x) = 1$.

5. Given the definitions of a function f:
- determine a formula that defines $f \circ f$;
- determine the domain of $f \circ f$;
- evaluate $(f \circ f)(4)$, $(f \circ f)(0)$, and $(f \circ f)(-3)$.

A. $f(x) = 5x$

F. $f(x) = 8x - 3$

B. $f(x) = -x/2$

G. $f(x) = 9/x$

C. $f(x) = \sqrt{x}$

H. $f(x) = \sqrt{3x - 7}$

D. $f(x) = \sqrt{4 - 6x}$

I. $f(x) = \sqrt[3]{x}$

E. $f(x) = (x + 1)/(x - 6)$

J. $f(x) = -10/(x + 4)$

6. Given the definition of F, define functions f and g so that $F = f \circ g$.

A. $F(x) = \sqrt[4]{11x + 6}$

E. $F(x) = (1 - x)^3$

B. $F(x) = (x + 3)^2 + 6(x + 3) + 9$

F. $F(x) = \sqrt[3]{x^2 - x - 8}$

C. $F(x) = |x - 10|$

G. $F(x) = |3x|^5$

D. $F(x) = 1/(2x - 5)^2$

H. $F(x) = (x + 2)/[4(x + 2) + 13]$

7. Given definitions of f, g, and h, determine a formula that defines $f \circ g \circ h$.

A. $f(x) = 2x$;
$g(x) = 3x$;
$h(x) = 5x$

C. $f(x) = x - 1$;
$g(x) = x - 4$;
$h(x) = x - 9$

B. $f(x) = x/4$;
$g(x) = x^2 - 5$;
$h(x) = \sqrt{4x + 1}$

D. $f(x) = \sqrt{-7 + x/2}$;
$g(x) = 2x^2 - 12x + 32$;
$h(x) = x + 3$

8. Suppose f and g are functions. Prove each statement.

A. If f and g are even, then $f + g$ is even.

B. If f and g are odd, then fg is even.

C. If f is odd and g is even, then f/g is odd.

D. If f is even and g is odd, then $f \circ g$ is even.

Set 91: Inverses of Functions

1. Suppose f and g are one-to-one functions where:

$$f(-4) = 6 \qquad f(0) = 1 \qquad f(1) = -2 \qquad f(3) = 0 \qquad f(6) = 5$$
$$g(-4) = 7 \qquad g(-2) = 6 \qquad g(0) = 3 \qquad g(3) = 1 \qquad g(5) = -4$$

Evaluate each expression.

A. $f^{-1}(-2)$

B. $g^{-1}(7)$

C. $(fg^{-1})(6)$

D. $(f^{-1} \circ g)(3)$

E. $(g^{-1} \circ f^{-1})(5)$

F. $(f^{-1} \circ (g - f))(-4)$

G. $g^{-1}(1)$

H. $f^{-1}(0)$

I. $(g/f^{-1})(5)$

J. $(f^{-1} \circ g^{-1})(-4)$

K. $(f \circ g^{-1})(1)$

L. $(g^{-1} \circ (g + f^{-1}))(-2)$

2. Verify that the given functions are inverses of each other.

A. $f(x) = -x/7$;
   $g(x) = -7x$

B. $f(x) = 5x - 3$;
   $g(x) = (x + 3)/5$

C. $f(x) = (\sqrt[3]{x - 1})/2$;
   $g(x) = 1 + 8x^3$;

D. $f(x) = x^2 - 2x + 5$ for $x \geq 1$;
   $g(x) = 1 + \sqrt{x - 4}$

E. $f(x) = 8x$;
   $g(x) = x/8$

F. $f(x) = (-x - 4)/7$;
   $g(x) = -4 - 7x$

G. $f(x) = x^5 - 9$;
   $g(x) = \sqrt[5]{x + 9}$

H. $f(x) = -2 + \sqrt{x + 14}$;
   $g(x) = x^2 + 4x - 10$ for $x \geq -2$

3. Provide a counterexample to show that the given functions are not inverses of each other.

A. $f(x) = 3x + 6$;
   $g(x) = -1 + x/3$

B. $f(x) = \sqrt{x + 5}$;
   $g(x) = x^2 - 5$

C. $f(x) = x - 8$;
   $g(x) = x + 9$

D. $f(x) = x$;
   $g(x) = |x|$

4. Fill in the blanks.

    A. If f and g are inverses and $f(5) = -2$, then $g(-2) =$ _____ .

    B. If f and g are inverses and $g(3) = 0$, then $f(0) =$ _____ .

    C. If f is a one-to-one function with range $\mathbf{R}$, then $(f \circ f^{-1})(7) =$ _____ .

    D. If g is a one-to-one function with domain $\mathbf{R}$, then $g^{-1}(g(-2)) =$ _____ .

5. Determine the formula and domain that define the inverse of the given function.

    A. $f(x) = 2x - 5$ I. $h(x) = 10 - x$

    B. $g(x) = \sqrt{3x + 8}$ J. $f(x) = \sqrt{1 - 6x}$

    C. $h(x) = 32x^5 - 11$ K. $g(x) = 10 + 27x^3$

    D. $f(x) = \sqrt[3]{12x + 5}$ L. $h(x) = 2\sqrt[5]{3 - x}$

    E. $g(x) = -x^2 - 7$ for $x \geq 0$ M. $f(x) = (x + 5)^2 + 8$ for $x \leq -5$

    F. $h(x) = (3x + 4)/(1 - 9x)$ N. $g(x) = -x$

    G. $f(x) = 1/x$ O. $h(x) = x/(11 + 6x)$

    H. $g(x) = x^2 - 8x - 2$ for $x \leq 4$ P. $f(x) = 2x^2 + 4x + 1$ for $x \geq -1$

6. Suppose f and g are one-to-one functions. Prove that $(f \circ g)^{-1} = g^{-1} \circ f^{-1}$.

Set 92: Linear Functions

1. Draw the graph of each linear function.

    A. $f(x) = x + 5$ D. $g(x) = 2x - 8$

    B. $h(x) = 4 - 3x$ E. $f(x) = -x/4 - 1$

    C. $g(x) = 2$ F. $h(x) = -7$

2. Define a linear function f that satisfies the given conditions.

    A. the graph of f passes through the points $(-4, 2)$ and $(-2, 8)$

    B. the graph of f passes through the points $(5, -9)$ and $(0, -9)$

    C. $f(-1) = 8$ and $f(3) = 0$

    D. $f(2) = 3$ and $f(6) = 13$

    E. the graph of f has x-intercept 7 and y-intercept 5

    F. the graph of f has x-intercept c and y-intercept c and $f(4) = -12$
    for some $c \in \mathbf{R}$

3. Suppose f is a linear function whose graph has nonzero slope. Prove that f^{-1} is a linear function.

4. Suppose f and g are linear functions. Prove that $f + g$ is a linear function.

5. Suppose $f(x) = 5x - 3$ and $g(x) = 2x + c$ for some constant $c \in \mathbf{R}$. Find the value of c for which $(f \circ g)(x) = (g \circ f)(x)$.

Set 93: Quadratic Functions

1. For each quadratic function:
   - determine the vertex and axis of symmetry of its graph;
   - determine its extremum;
   - draw its graph.

   A. $f(x) = 2(x - 7)^2 + 3$ G. $g(x) = -3(x + 1)^2 - 8$

   B. $h(x) = -x^2 - 10$ H. $f(x) = (1/2)x^2$

   C. $g(x) = x^2 + 4x - 1$ I. $h(x) = 2x^2 + 2x - 9$

   D. $f(x) = 3x^2 - 18x + 20$ J. $g(x) = x^2 - 10x + 2$

   E. $h(x) = -(1/2)x^2 - 7x$ K. $f(x) = -x^2 + 5x + 1$

   F. $g(x) = 4x^2 - 3x - 6$ L. $h(x) = 3x^2 - 2x + 6$

2. Define a quadratic function f whose graph satisfies the given conditions. Leave the formula in the form $f(x) = a(x - h)^2 + k$.

   A. vertex $(2, -11)$; passes through $(-1, 16)$

   B. vertex $(-5, 3)$; passes through $(-3, -1)$

   C. axis of symmetry $x = -1$; passes through $(1, -6)$ and $(-5, 0)$

   D. axis of symmetry $x = 4$; passes through $(5, -1)$ and $(2, -7)$

   E. passes through $(-2, 5)$, $(-1, 0)$, and $(0, -7)$

3. Solve each word problem.

   A. A boy atop a 150-foot cliff throws a stone at an eagle. The stone misses the bird and falls to the ground below. The height (in feet) of the stone above the ground t seconds after it leaves the boy's hand is given by the formula $h(t) = -16t^2 + 40t + 150$. Find the maximum height that the stone attains and the time at which it attains this height.

   B. A rancher wants to build a rectangular pen along a river using 200 yards of fencing. Assuming that the river serves as one side of the pen and that the rancher requires fencing only on the other three sides, what dimensions will maximize the area of the pen?

C. Find the pair of numbers with the largest product where the sum of the numbers is 30.

D. For what two numbers is the sum of the squares of those numbers the smallest if the sum of the numbers is 10?

E. The profit (in dollars) that a company earns on production of x widgets is given by the function $P(x) = -3x^2 + 300x + 20000$. Determine the maximum profit of the company and the level of production at which it attains this maximum profit.

F. The usual price of entrance to an amusement park is $18; however, customers who visit the park in a group all receive a discount of $0.20 for each additional person in the group beyond forty people. What size of group provides maximum revenue to the owner of the park?

G. Suppose $x + y = 6$. Find the maximum value of $y^2 - 3x^2$.

H. Find the point on the line $2x - y = 3$ that lies nearest the point $(4, 0)$. [Hint: Note that the distance to a point is minimized where the square of the distance to the point is minimized.]

Set 94: The Remainder and Factor Theorems

1. For each polynomial $f(x)$ and constant c:
- determine $f(c)$ using the Remainder Theorem and
- evaluate $f(c)$ directly.

A. $f(x) = x^5 + x^4 - 7x^3 - 2x^2 + 9x - 5;\ c = 2$

B. $f(x) = x^6 + 4x^5 + 5x^3 - 3x^2 - 2x + 2;\ c = -1$

C. $f(x) = 2x^4 + 4x^3 - x^2 + 12x + 4;\ c = -3$

D. $f(x) = -3x^3 + 20x^2 - 13x + 7;\ c = 6$

2. Use the Factor Theorem to determine whether $g(x)$ is a factor of $f(x)$.

A. $f(x) = 2x^4 - 5x^3 + 7x^2 + 5x - 9;\ g(x) = x - 1$

B. $f(x) = x^3 - 2x^2 - 7x - 5;\ g(x) = x - 4$

C. $f(x) = x^3 + 8x^2 + 5x - 10;\ g(x) = x + 2$

D. $f(x) = 3x^5 + 13x^4 + 6x^3 + 9x^2 - 16;\ x + 4$

3. Factorize each polynomial completely.

A. $f(x) = 2x^2 + 7x - 1$

E. $g(x) = 3x^2 - 9x + 4$

B. $g(x) = x^3 - 6x^2 - x + 30;$
[Hint: $g(3) = 0$]

F. $h(x) = x^3 + 13x^2 + 47x + 35;$
[Hint: $h(-1) = 0$]

C. $h(x) = 3x^3 + 19x^2 + 8x - 80;$
[Hint: $h(-4) = 0$]

G. $f(x) = 2x^3 + 4x^2 - 58x + 84;$
[Hint: $f(2) = 0$]

D. $f(x) = 3x^3 - 25x^2 + 7x + 8;$
[Hint: $f(8) = 0$]

H. $g(x) = 2x^3 + 14x^2 + 17x - 15;$
[Hint: $g(-5) = 0$]

4. Show that the polynomial $5x^4 + 8x^2 + 1$ has no factors of the form $x - c$, where $c \in \mathbf{R}$ is a constant.

5. The remainder yielded by dividing a certain polynomial $f(x)$ by $x - 2$ is 6 and the remainder yielded by dividing $f(x)$ by $x + 3$ is 11. Determine the remainder yielded by dividing $f(x)$ by $(x - 2)(x + 3)$.

6. Suppose $9x^3 + 6x^2 - 20x + 8$ is divisible by $(x - c)^2$ where $c > 0$. Find c.

Set 95: The Rational Zeros Theorem

1. Determine the set of all possible rational zeros of the given polynomial.

A. $4x^3 - 5x^2 + x - 2$

C. $3x^4 + 2x^3 - 9x^2 - 11x - 6$

B. $x^3 - 3x^2 + 2x^2 - 7x + 5$

D. $8x^5 - 3x^2 - 5x - 1$

2. For each polynomial:
- determine one rational zero of the polynomial and
- factorize the polynomial completely.

A. $x^3 + 12x^2 - 38x - 21$

D. $x^3 - x^2 - 11x - 10$

B. $2x^3 - 7x^2 - x + 15$

E. $3x^3 - 4x^2 - 29x + 8$

C. $7x^4 - 26x^3 - 85x^2 - 64x - 12$

F. $2x^4 + 11x^3 - 78x^2 + 116x - 40$

Set 96: Descartes' Rule of Signs

1. For each polynomial $f(x)$, determine:
   - the numbers of variations in sign within $f(x)$ and $f(-x)$ and
   - the possible numbers of positive and negative real zeros of $f(x)$.

A. $f(x) = -x^3 + 4x^2 - 3x + 5$

E. $f(x) = 2x^4 - x^3 - 6x^2 - 8x + 11$

B. $f(x) = 8x^2 + 4x + 13$

F. $f(x) = -3x^5 - 2x^4 + x^2 + 9x + 1$

C. $f(x) = 16x^4 + 1$

G. $f(x) = -27x^3 - 8$

D. $f(x) = 7x^5 - 2x^2 + 8x^3 - 10$

H. $f(x) = 5x^2 + 3 - 2x$

I. $f(x) = 2x^7 - 5x^6 + 11x^5 - 3x^4 - 2x^3 + 4x^2 - 12x + 7$

Set 97: Dominant Terms of Polynomials

1. For each polynomial:
   - state its dominant term and
   - determine its behavior as $x \to -\infty$ and as $x \to \infty$.

A. $f(x) = -x^3 + 4x^2 - 3x + 5$

F. $f(x) = 2x^4 - x^3 - 6x^2 - 8x + 11$

B. $f(x) = 7x^5 - 2x^2 + 8x^3 - 10$

G. $f(x) = 7 - 25x^9$

C. $f(x) = 9 - 2x - 7x^2 - 4x^3$

H. $f(x) = 11x - 1 + 5x^3 - 3x^2$

D. $f(x) = -x(x+2)(x-3)(x-8)$

I. $f(x) = 3x(x-5)(x+7)$

E. $f(x) = 8x(1-x)(6-x)(2+7x)$

J. $f(x) = x(3-2x)(5+9x^2-22)$

Set 98: Polynomial Functions

1. For each polynomial function:
   - determine the x- and y-intercepts of its graph,
   - make its sign chart,
   - determine its behavior as $x \to -\infty$ and as $x \to \infty$, and
   - use the intercepts, sign chart, and behavior to sketch its graph crudely.

A. $f(x) = (x+2)(x-1)(x-3)$

B. $g(x) = \frac{1}{2}x^4(x+4)(7-x)$

C. $h(x) = -7x(x-3)$

D. $f(x) = x^3 - 2x^2 - 9x + 18$

E. $g(x) = -2x^3 + 14x$

F. $h(x) = 5x^3 - 47x^2 + 50x + 48$
   [Hint: $h(2) = 0$]

G. $h(x) = -3x(x+5)(x-2)(x-4)$

H. $f(x) = (x-3)^2(x+1)(x-8)$

I. $g(x) = \frac{1}{8}(5-2x)(10-x)$

J. $h(x) = -x^3 + 10x^2 - 25x$

K. $f(x) = (x+6)(x-1)(x-2) - 60$

L. $g(x) = -4x^3 + 16x^2 + 35x - 147$
   [Hint: $g(-3) = 0$]

Set 99: Asymptotes

1. Find the equations of the asymptotes of the graph of each function.

A. $r(x) = -\dfrac{7}{x+5}$

G. $r(x) = \dfrac{20}{(7x-4)(2x+8)}$

B. $r(x) = \dfrac{x^2 - x + 3}{2x^2 - 17x - 9}$

H. $r(x) = \dfrac{3x^2 + 11x}{12x^2 + 22x + 10}$

C. $r(x) = \dfrac{-8x^2 - 15x + 23}{x+3}$

I. $r(x) = \dfrac{2x^2 - 5x - 17}{x - 5}$

D. $r(x) = \dfrac{2x^2 + 14x + 22}{2x+4}$

J. $r(x) = \dfrac{-10x^3 + 7x^2 + 25x - 9}{2x^2 - x - 6}$

E. $r(x) = \dfrac{2x^2 - 5x - 7}{6x^2 - 17x - 14}$

K. $r(x) = \dfrac{x^3 - 3x^2 + 5x - 15}{x^2 - 9}$

F. $r(x) = \dfrac{26 - 6x}{3x - 13}$

L. $r(x) = \dfrac{x^2 + 3x - 10}{x+5}$

Set 100: Rational Functions

1. For each rational function:
   - determine the x- and y-intercepts of its graph,
   - make its sign chart,
   - find the equations of the asymptotes of its graph, and
   - use the intercepts, chart, and asymptotes to sketch its graph crudely.

A. $r(x) = -\dfrac{12}{x+4}$

B. $r(x) = \dfrac{36}{(x-3)^2(4x-1)^2}$

C. $r(x) = \dfrac{x^2-4x+3}{2x^2-17x-9}$

D. $r(x) = \dfrac{-8x^2-14x+22}{x+3}$

E. $r(x) = \dfrac{2x^2+14x+12}{2x+4}$

F. $r(x) = \dfrac{2x^2-5x-7}{6x^2-17x-14}$

G. $r(x) = \dfrac{8-6x}{3x-4}$

H. $r(x) = \dfrac{40}{(7x-4)(2x+10)}$

I. $r(x) = -\dfrac{8}{(x+2)^2(7x+3)}$

J. $r(x) = \dfrac{3x^2+14x}{12x^2+46x+30}$

K. $r(x) = \dfrac{2x^2-5x-18}{x-5}$

L. $r(x) = \dfrac{-10x^3+7x^2+21x-18}{2x^2-x-6}$

M. $r(x) = \dfrac{x^3-3x^2+5x-15}{x^2-9}$

N. $r(x) = \dfrac{x^2+3x-10}{x+5}$

2. Suppose the population P (in millions) of a certain city at a time t years from now is given by the formula $P = (3t+1)/(8t+10)$. To what value will the population approach in the long run?

Set 101: Irrational Exponents

1. Approximate each expression to five decimal places using rational exponents and a calculator.

A. $6^{\sqrt{3}}$
[Hint: $\sqrt{3} \approx 1.732050807569$]

B. $10^{\sqrt{8}}$
[Hint: $\sqrt{8} \approx 2.828427124746$]

C. 4^{π}
[Hint: $\pi \approx 3.141592653590$]

D. $7^{\sqrt{5}}$
[Hint: $\sqrt{5} \approx 2.236067977500$]

E. $2^{\sqrt{29}}$
[Hint: $\sqrt{29} \approx 5.385164807135$]

F. 9^{e}
[Hint: $e \approx 2.718281828459$]

Set 102: Exponential Functions

1. Solve each equation.

A. $6^x = 36$

B. $3^{10x} = 27$

C. $25^{x+3} = \sqrt{5}$

D. $4^{5x-7} = 8^{2x+4}$

E. $11^{3x+8} \cdot 11^{2x-3} = 1$

F. $2^{x^2}/2^{7x} = 8^6$

G. $243 \cdot 3^{4x} = 27^{x+2}$

H. $5^x = 0$

I. $x^2(4^x) - 10x(4^x) + 21(4^x) = 0$

J. $2^x = 1/32$

K. $5^{3x-1} = 25$

L. $12^{9x} = 24\sqrt{3}$

M. $81^{2-x} = (1/27)^{4x-7}$

N. $16^{7x+12}/16^{6x+10} = 1/2$

O. $9^{x^2} \cdot 9^{2x} = 3^{30}$

P. $8 \cdot 2^{x+13} = 4^{2x-7}$

Q. $7^x = -49$

R. $3x^2(13^x) = 12x(13^x)$

2. Determine the value of b for which the graph of the function $f(x) = b^x$ passes through the given point.

A. $(4, 625)$

B. $(-3, 1/8)$

C. $(5, 1/100{,}000)$

D. $(-2, 1/81)$

E. $(2, 121)$

F. $(6, 1/64)$

3. For each function:
- determine its domain and range,
- find the equation of the asymptote of its graph, and
- draw its graph.

A. $f(x) = 3^x$

B. $g(x) = (1/4)^{-x}$

C. $h(x) = 2^{x-3} + 1$

D. $f(x) = -(1/5)^{x+6} - 2$

E. $g(x) = (1/3)^x$

F. $h(x) = 4^{-x}$

G. $f(x) = -5^{x+2} + 11$

H. $g(x) = 4 \cdot (1/2)^x - 5$

Set 103: Logarithmic Functions

1. Evaluate each logarithm.

A. $\log_3 81$

I. $\log 1{,}000{,}000$

B. $\log_{32} 2$

J. $\log_{125} 5$

C. $\log_{12} 1/144$

K. $\log_7 1/343$

D. $\log_{1/4} 64$

L. $\log_{1/9} 81$

E. $\log_{13} 13^8$

M. $\log_4 16^{41}$

F. $\log_{17} \sqrt{17}$

N. $\log_2 8\sqrt{2}$

G. $\log_8 1$

O. $\log_{97} 97$

H. $\ln e$

P. $\log_{872} 1$

2. Write each equation in exponential form.

A. $\log_2 8 = 3$

F. $\log_5 625 = 4$

B. $\ln e^9 = 9$

G. $\log 100 = 2$

C. $\log_{16} 4 = 1/2$

H. $\log_9 1/81 = -2$

D. $\log_{1/7} 7 = -1$

I. $\log_x 11 = -4$

E. $\log_{19} x = 80$

J. $\log_{243} 1/3 = -1/5$

3. Write each equation in logarithmic form.

A. $7^3 = 343$

F. $3^5 = 243$

B. $10^4 = 10{,}000$

G. $e^6 = e^6$

C. $32^{1/5} = 2$

H. $36^{-1/2} = 1/6$

D. $9^{-1} = 1/9$

I. $(1/4)^{-4} = 256$

E. $x^8 = 50$

J. $5^x = 212$

4. Determine the domain of the given function.

A. $f(x) = \log_3(4x + 7)$

D. $g(x) = \log_5(9 - 2x)$

B. $h(x) = \log_8(x^2 - 5x - 24)$

E. $f(x) = \log_2 x^2$

C. $g(x) = \log_{10}(|x| + 3)$

F. $h(x) = \log_9(x^3 + 11x^2 + 30x)$

5. For each function:
- determine its domain and range,
- find the equation of the asymptote of its graph, and
- draw its graph.

A. $f(x) = \log_3 x$

E. $g(x) = \log_{1/3} x$

B. $g(x) = \log_{1/4}(-x)$

F. $h(x) = \log_4(-x)$

C. $h(x) = 1 + \log_2(x - 3)$

G. $f(x) = 11 - \log_5(x + 2)$

D. $f(x) = -2 - \log_{1/5}(x + 6)$

H. $g(x) = -5 + 4\log_{1/2} x$

6. Find the value that $\log_2(3x + 4) - \log_2(24x - 7)$ approaches as $x \to \infty$.

Set 104: Properties of Logarithms

1. Write each expression as a sum and/or difference of logarithms with no exponents. Simplify where possible.

A. $\log_2(13/x)$

F. $\log_9 4x$

B. $\log_5 8x^{25}$

G. $\log_7 49^m n$

C. $\log 1000x^5 y$

H. $\log_2(a^2 b^7 /32)$

D. $\log_6 \sqrt{x^8 y^2 /z^3}$

I. $\ln \sqrt[3]{1/(x^9 y^{15})}$

E. $\log_3 \sqrt[8]{81(m-4)^4(n+3)^2 p^5}$

J. $\log_5 \sqrt[4]{x^2(3y^2+10)^{12}/125}$

2. Write each expression as single logarithm. Simplify where possible.

A. $\log_7 x + \log_7 y$

F. $\log_8 w - \log_8 11$

B. $5\log_{11} x - 2\log_{11} y$

G. $2\log 9 + 8\log m - 3\log n$

C. $\log_3 45 - \log_3 5$

H. $\log_6 9 + \log_6 24$

D. $\ln(x+5) + \ln(3x-7)$

I. $\log_4 x + \log_4 23 + \log_4(9x+1)$

E. $\log_2(x^2 - 16) - \log_2(x-4)$

J. $\log_5(x^2 + x - 6) - \log_5(x+3)$

3. Suppose $a = \log 2$, $b = \log 3$, and $c = \log 5$. Write each logarithm in terms of a, b, and c.

A. $\log 15$

D. $\log 0.4$

B. $\log 27/4$

E. $\log 60$

C. $\log_6 50$

F. $\log_8 4.5$

4. Use the Change of Base Formula to rewrite each expression in terms of natural logarithms, common logarithms, and logarithms with base 2.

A. $\log_9 71$

E. $\log_5 43$

B. $\log_7 100$

F. $\log_{10} 63$

C. $\log_4 55$

G. $\log_{11} 128$

D. $\log_k e$

H. $\log_a b$

5. Use the Change of Base Formula to rewrite each expression in terms of natural logarithms, then use a calculator to approximate the value of the expression to five decimal places.

A. $\log_8 111$

E. $\log_{21} 5$

B. $\log_{486} 3$

F. $\log_3 250$

C. $\log_{11} 0.00365$

G. $\log_6 0.0000292$

D. $\log 0.714$

H. $\log 969$

6. Simplify each expression.

A. $\log 11 \div \log(1/11)$

D. $\log 9 \div \log 3$

B. $\log 2 \div \log 32$

E. $\log(4/7) \div \log(7/4)$

C. $\log_4[(\log_{81} 3)^{(\log_2 32)}]$

F. $\log_9[(\log_8 2)^{(\log_5 625)}]$

G. $(\log_4 5)(\log_5 6)(\log_6 7) \cdots (\log_{62} 63)(\log_{63} 64)$

H. $\dfrac{8}{\log_2 210} + \dfrac{8}{\log_3 210} + \dfrac{8}{\log_5 210} + \dfrac{8}{\log_7 210}$

7. Determine x in terms of a, b, and c where:
- $(\log u)/a = (\log v)/b = (\log w)/c = \log y$;
- $vw^3/u^2 = y^x$;
- $y \neq 1$.

Set 105: Exponential Equations

1. Solve each equation. Leave your solutions in terms of natural or common logarithms.

A. $2^x = 77$

M. $5^x = 2000$

B. $10^x = 589/3$

N. $e^x = 17/99$

C. $9^x = -4$

O. $3^x = 0$

D. $11^{4x+5} = 3$

P. $7^{16-3x} = 0.23$

E. $3^{x^2} = 91$

Q. $4^{|x|} = 85$

F. $14 \cdot 6^{-x} + 7 = 9$

R. $2^x / 2^{9-4x} = 38$

G. $5^{8x+2} \cdot 5^{3x-6} = 347$

S. $12 \cdot 7^{4x} - 3 = 5$

H. $x^2 \cdot 8^x - 3x \cdot 8^x - 28 \cdot 8^x = 0$

T. $3x^2 \cdot 3^x + 17x \cdot 3^x + 10 \cdot 3^x = 0$

I. $2 \cdot 7^{2x} - 19 \cdot 7^x + 9 = 0$

U. $5^{2x} - 10 \cdot 5^x - 24 = 0$

J. $10^x - 3 \cdot 10^{-x} = 2$

V. $e^x + 6e^{-x} = 5$

K. $13^{5x+8} = 6^{3x-2}$

W. $9^{10x-3} = 4^{11x-8}$

L. $e^{6x+1} = 10^{2x+9}$

X. $10^{11-x} = e^{7-6x}$

2. Solve each system of equations.

A. $\begin{cases} 2^{y-7} = 8^x \\ 3^{x+9} = 9^y \end{cases}$

B. $\begin{cases} 32^{2-y} = 4^{x+3} \\ 25^{x-1} = 125^{6+y} \end{cases}$

Set 106: Logarithmic Equations

1. Solve each equation.

A. $\log_5 x = 3$

L. $\log_3 x = 4$

B. $\log_7(3x - 8) = 2$

M. $\log_2(x^2 + 7) = 5$

C. $\log_x 1/27 = -3$

N. $\log_{5x} 11 = 1/2$

D. $\log_{2x+16} 64 = 6$

O. $\log_{9-4x} 81 = 4$

E. $\log_6 10 + \log_6(x + 13) = 2$

P. $\log_4(x - 5) + \log_4 10 = 3$

F. $\log_3(x - 5) + \log_3(x + 3) = 2$

Q. $\log_6(x + 4) + \log_6(x - 1) = 2$

G. $\ln(2x - 9) + \ln(x - 5) = \ln 3$

R. $\log(x - 2) + \log(3x - 7) = \log 10$

H. $\log_{11}(13x + 6) - \log_{11}(x + 2) = 1$

S. $\log_2(11x - 13) - \log_2(x + 1) = 3$

I. $2\log x = \log(12 - x)$

T. $2\ln x + \ln 2 = \ln(11x - 15)$

J. $\log x^2 = (\log x)^2$

U. $\ln \sqrt[3]{x} = \sqrt[3]{\ln x}$

K. $\dfrac{1}{\log_4 x} + \dfrac{1}{\log_5 x} + \dfrac{1}{\log_{50} x} = 3$

V. $\dfrac{3}{\log_4 x} + \dfrac{3}{\log_6 x} + \dfrac{3}{\log_{54} x} = 12$

2. Solve each system of equations. Assume $x, y \in (0, \infty)$.

A. $\begin{cases} x^y = y^x \\ y = 4x \end{cases}$

B. $\begin{cases} \log_x y + \log_y x = 17/4 \\ xy = 7 \end{cases}$

Set 107: Exponential Inequalities

1. Solve each inequality. Leave your solutions in terms of natural or common logarithms.

A. $2^x > 39$

B. $10^{4x+9} < 0$

C. $(0.7)^{2x+5} \leq 11$

D. $5(6^{-x-3} + 7) > 75$

E. $5^{3x-4} \geq 21^x$

F. $10^{x^2-4x} \leq 10^{12}$

G. $5^{8x-1} < 4$

H. $8^{x+7} > 0$

I. $(0.1)^{9-3x} \geq 0.001$

J. $4(11 - 3^{(40-x)/10}) < 28$

K. $(1/3)^{1-2x} \leq 17^{-x}$

L. $e^6 \geq e^{-7x-x^2}$

2. Determine the largest natural number m such that $m^{278} < 5^{417}$. Do not use a calculator.

Set 108: Logarithmic Inequalities

1. Solve each inequality.

A. $\log_5(x - 8) < 2$

G. $\log_2(3x - 15) < 3$

B. $\log_{0.3}(16 - 2x) \geq 4$

H. $\log_{0.2}(5x + 10) \leq 4$

C. $\ln(x^2 - 6x - 16) > \ln 24$

I. $\log(x^2 + 16x + 63) < \log 8$

D. $\log_2(2x^2 - 18) \leq 5$

J. $\log_{13}(x^2 - 36) \geq 1$

E. $\log_3 x + \log_3(x - 6) < 3$

K. $\log_2 x + \log_2(x + 6) \geq 4$

F. $\log_{0.5}(x + 7) - \log_{0.5}(2x - 11) > 0$

L. $\log_{0.7}(3x - 8) - \log_{0.7} 100 \geq 2$

Set 109: Compound Interest

1. A woman places $500 in a savings account at a bank that pays interest at an annual rate of 6%. Determine the amount of money in her account after ten years for each frequency of compounding.

   A. weekly

   B. monthly

   C. semiannually

   D. daily

   E. annually

   F. quarterly

   G. biweekly

   H. continuously

2. A man invests $10,000 in a fund on which interest is compounded monthly. Determine the amount of money in the fund after twenty years for each annual rate of interest.

   A. 9%

   B. 2%

   C. 5%

   D. 1%

   E. 6%

   F. 18%

3. Solve each word problem.

   A. A man bought a certificate of deposit for $1,000 on the first day of 1995. He earned interest on the CD at an annual rate of 3% compounded semiannually. In what year had the value of the CD grown to $5,500.

   B. A worker deposited $50 in an account on the first day of 2003. He earned interest on the money in the account at an annual rate of 5% compounded continuously. In what year had the amount of money in the account grown to $275.

   C. How long will it take an investment to double in value if it bears interest at an annual rate of 4% compounded monthly.

   D. How long will it take an investment to triple in value if it bears interest at an annual rate of 12% compounded quarterly.

   E. A gold coin appreciated in value from $350 in 2002 to $1,875 in 2019. At what annual rate did the coin appreciate?

F. Suppose a new sedan depreciates in value from $25,000 to $7,000 in five years. At what annual rate does the sedan depreciate?

G. A magnate lent $2,000 to his needy neighbor. If interest on the loan is compounded continously and the borrower must pay the lender $3,000 in five years to extinguish the debt, then what was the annual rate of interest on the loan?

H. The winner of a lottery placed $500,000 in a savings account on which interest is compounded weekly. Seven years later, the amount of money in the account had grown to $800,000. What annual rate of interest did the winner earn on his deposit?

I. A 45-year old man wants to have $1,000,000 when he retires in thirty years. How much money should he place now in a account that pays interest at an annual rate of 2% compounded monthly to accomplish his goal?

J. A 12-year old boy would like to have $5,000,000 when he retires in fifty years. How much of the money he has saved from wages on his paper route must he place in a certificate of deposit that earns interest at an annual rate of 17.2% compounded daily in order to accomplish his goal?

K. Which investment has the greater value after ten years:
 - a deposit of $100 that earns 5% compounded continuously, or
 - a deposit of $110 that earns 4% compounded continuously?

L. Which investment provides the greater return:
 - one that earns 6% compounded monthly, or
 - one that earns 6.15% compounded annually?

4. Use the Rule of 72 to estimate the number of years required for an investment to double in value at the given annual rate of interest.

A. 2%

B. 18%

C. 9%

D. 6%

Set 110: Exponential Growth and Decay

1. Suppose a colony of Peruvian gnats grows exponentially according to the formula $Q_t = 34e^{0.051t}$, where Q_t represents the population of the gnats (in thousands) t days after April 1. Determine each of the following:

  A. the initial population of gnats on April 1;

  B. the constant of growth for the population;

  C. the population on April 10.

2. A substance decays exponentially according to the formula $Q_t = 981e^{-0.18t}$, where Q_t represents the amount of the substance (in ounces) t hours after 6:00 AM. Determine each of the following:

  A. the initial amount of the substance;

  B. the constant of growth for the population;

  C. the amount of the substance at 8:00 PM.

3. The constant of exponential decay for a certain substance is -0.072, where time is measured in months.

  A. Determine the half-life of the substance.

  B. What percent of the substance will remain after 50 months?

  C. How long will it take for the substance to decay to 15% of its original amount?

4. The constant of exponential growth for a certain kind of mold is 0.295, where time is measured in days.

  A. Determine the doubling time of the mold.

  B. By what multiple will the mold have grown after three weeks?

  C. How long will it take for the mold to grow to ten times its original amount?

5. The initial mass of a substance that decays exponentially is 200 grams. After 30 minutes, 156 grams of the substance remain. If the mass of the substance continues to decay at the same exponential rate, then after how many minutes will only 2 grams of the substance remain?

6. The initial volume of a blob that grows exponentially is 4 cubic inches. After two days, the volume of the blob is 7 cubic inches. If the volume of the blob continues to grow at the same exponential rate, then after how many days will its volume reach 50,000 cubic inches?

7. Suppose the initial population of a certain mining town in Alaska is 250 and the doubling time of the population is 61 days.

    A. Determine a formula for the population after t months.

    B. How long will it take the population to grow to 10,000?

    C. What will be the population of the town in two years?

8. Suppose a nurse injects a patient with 80 milligrams of a certain drug. Every six hours, the amount of the drug in the bloodstream of the patient falls by 10%.

    A. Determine a formula for the amount of the drug in the bloodstream after t hours.

    B. How long will it take the amount in the bloodstream to fall to 3% of its initial level?

    C. What amount of the drug will remain the bloodstream after 72 hours?

Set 111: Periodic Functions

1. Find the period p and amplitude a of the function with the given graph.

A.

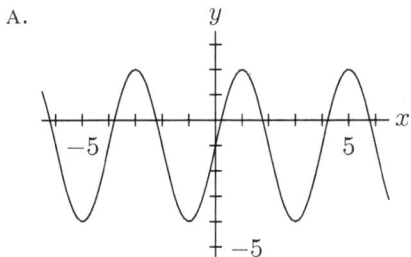

C.

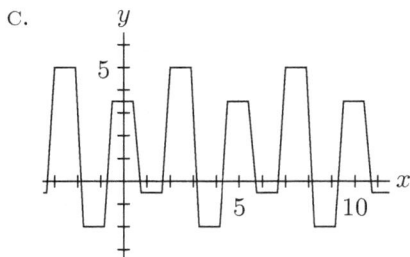

B.

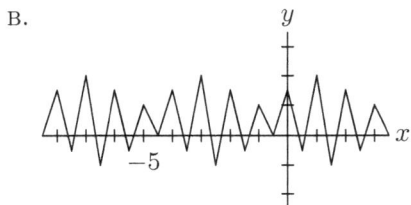

D.

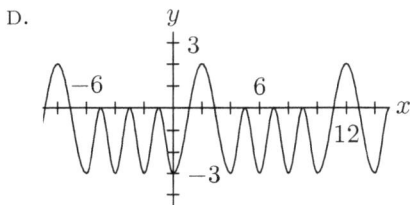

2. Determine the phase shift of the function g with reference to f, assuming f is a periodic function.

A. $g(x) = f(x - 5)$

C. $g(x) = f(x + 8)$

B. $g(x) = f(x + \pi)$

D. $g(x) = f(x - e)$

3. Determine the period p_g and amplitude a_g of the function g, assuming f is a periodic function with the given period p_f and amplitude a_f.

A. $p_f = 4$,
$a_f = 7$,
$g(x) = -3 \cdot f(2x - 12)$

C. $p_f = 9$,
$a_f = 2$,
$g(x) = 4 \cdot f(-3x + 15)$

B. $p_f = 2\pi$,
$a_f = 1$,
$g(x) = 8 \cdot f(\pi x) - 7$

D. $p_f = \pi$,
$a_f = 10$,
$g(x) = -5 \cdot f(7(x - 3)/4) + 1$

Set 112: Circles

1. Determine the equation of the circle C with the given characteristics. Leave the equation in standard form.

A. center: $(3, 8)$; radius: 7

F. center: $(-2, 6)$; radius: 9

B. center: $(0, -7)$; radius: 1

G. center: $(8, 0)$; radius: 4

C. center: $(-1, -2)$;
  the point $(3, -5)$ lies on C

H. center: $(4, 11)$;
  the point $(-1, 11)$ lies on C

D. center: $(-3, 7)$;
  the point $(-8, 9)$ lies on C

I. center: $(-4, 9)$;
  the point $(4, -6)$ lies on C

E. center: $(2, -3)$;
  the x-axis lies tangent to C

J. center: $(-8, -5)$;
  the y-axis lies tangent to C

K. points $(-4, -5)$ and $(8, 0)$ lie on C;
  the center of C lies on the line through points $(-4, -5)$ and $(8, 0)$

L. points $(2, -7)$ and $(4, -3)$ lie on C;
  the center of C lies on the line through points $(2, -7)$ and $(4, -3)$

2. Determine the center and the radius of the circle with the given equation and then draw the circle.

A. $(x - 8)^2 + (y + 1)^2 = 9$

E. $(x + 2)^2 + (y - 5)^2 = 16$

B. $(x - 3)^2 + (y - 4)^2 = 81$

F. $(x + 9)^2 + (y + 6)^2 = 49$

C. $x^2 + y^2 = 36$

G. $(x - 5)^2 + y^2 = 17$

D. $x^2 + (y + 7)^2 = 10$

H. $3(x - 4)^2 + 3(y - 10)^2 = 12$

3. Convert each equation of a circle into standard form and then identify the center and the radius of the circle.

A. $x^2 + 12x + y^2 - 10y = 20$

E. $x^2 - 16x + y^2 - 2y = -29$

B. $x^2 + y^2 + 20y + 36 = 0$

F. $x^2 + 14x + y^2 - 72 = 0$

C. $4x^2 - 16x + 4y^2 + 20y = 19$

G. $9x^2 + 48x + 9y^2 + 18y = -55$

D. $3x^2 - 10x + 3y^2 - 4y - 7 = 0$

H. $2x^2 + 11x + 2y^2 - 3y + 4 = 0$

4. Determine whether the point at each pair of coordinates lies on the circle with equation $(x + 7)^2 + (y - 4)^2 = 100$.

A. $(-1, -4)$ D. $(-7, 14)$

B. $(2, 8)$ E. $(-15, 10)$

C. $(3, 4)$ F. $(0, -3)$

5. Draw the graph of each inequality.

A. $(x - 2)^2 + (y - 7)^2 > 25$ C. $(x + 6)^2 + (y - 4)^2 < 64$

B. $(x + 3)^2 + y^2 \leq 4$ D. $x^2 + y^2 \geq 1$

6. Determine the area of the circle that passes through points $(4, 6)$, $(-3, 5)$, and $(-2, -2)$.

Set 113: Angles

1. For each angle, determine two labels that appropriately identify it.

A.

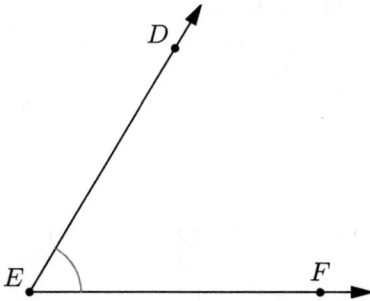

B.

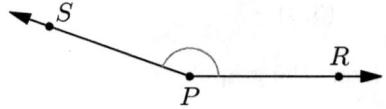

Set 114: Measures of Angles in Degrees

1. Draw in standard position the angle with the given measure.

   A. $120°$ G. $315°$

   B. $30°$ H. $240°$

   C. $270°$ I. $-225°$

   D. $-300°$ J. $-180°$

   E. $405°$ K. $660°$

   F. $-570°$ L. $-1125°$

2. Find measures, two positive and two negative, of four angles whose terminal sides coincide with that of the angle of given measure. Assume all angles lie in standard position.

   A. $210°$ C. $415°$

   B. $-560°$ D. $-58°$

3. The terminal side of the angle with measure θ coincides with the terminal side of the angle with the given measure when drawn in standard position. Find θ if $0° \leq \theta < 360°$.

   A. $1560°$ C. $1000°$

   B. $-1900°$ D. $-505°$

4. Convert each measure to the form $d° \, m' \, s''$. Use a calculator if necessary and round numbers of seconds to the nearest whole number.

   A. $55.2417°$ C. $-38.4022°$

   B. $16.8309°$ D. $429.1655°$

5. Convert each measure into degrees, into minutes, and into seconds. Round your answers for degrees and minutes to four decimal places.

   A. $43° \, 2' \, 37''$ C. $20° \, 45' \, 32''$

   B. $100° \, 39' \, 56''$ D. $-3° \, 59' \, 16''$

Set 115: Measures of Angles in Radians

1. Suppose r denotes the radius of a circle centered at the vertex of an angle of measure θ and s denotes the length of the arc that subtends the angle. Determine the value of the indicated variable.

    A. Find θ (where $\theta > 0$)
        if $r = 3$ ft and $s = 18$ ft.

    D. Find θ (where $\theta < 0$)
        if $r = 14\pi$ m and $s = 2\pi$ m.

    B. Find s if $\theta = 5\pi/4$
        and $r = 6$ cm.

    E. Find s if $\theta = 11\pi/2$
        and $r = 18$ yd.

    C. Find r if $\theta = 3\pi/5$
        and $s = 9\pi$ mm.

    F. Find r if $\theta = 15$ and $s = 7$ in.

2. Draw in standard position the angle with the given measure.

    A. $5\pi/4$

    G. $\pi/3$

    B. $11\pi/6$

    H. $3\pi/4$

    C. $3\pi/2$

    I. π

    D. $-7\pi/4$

    J. $-7\pi/6$

    E. $8\pi/3$

    K. $19\pi/4$

    F. $-17\pi/6$

    L. $-23\pi/6$

3. Find measures, two positive and two negative, of four angles whose terminal sides coincide with that of the angle of given measure. Assume all angles lie in standard position.

    A. $5\pi/3$

    C. $-5\pi/6$

    B. $-15\pi/4$

    D. $-11\pi/2$

4. The terminal side of the angle with measure θ coincides with the terminal side of the angle with the given measure when drawn in standard position. Find θ if $0 \le \theta < 2\pi$.

    A. $23\pi/4$

    D. $10\pi/3$

    B. $5\pi/2$

    E. 19π

    C. $-31\pi/6$

    F. $-16\pi/7$

Set 116: Conversion between Degrees and Radians

1. Convert each measure from degrees to radians.

A. $60°$ G. $315°$

B. $210°$ H. $150°$

C. $-240°$ I. $-330°$

D. $-45°$ J. $-225°$

E. $540°$ K. $-990°$

F. $-1305°$ L. $405°$

2. Convert each measure from radians to degrees.

A. $11\pi/6$ G. $2\pi/3$

B. $5\pi/4$ H. $\pi/12$

C. $-7\pi/4$ I. $-7\pi/6$

D. $-5\pi/6$ J. $-\pi/3$

E. -8π K. $3\pi/2$

F. $10\pi/3$ L. $-13\pi/4$

3. The figure at right shows a rough cross-section of the earth through its center (point C), the north pole (point N), and Winnipeg, Manitoba (point W). The latitude of Winnipeg is approximately 49.53°.

Find the distance along the earth between Winnipeg and the equator (point E) under the assumption that the earth is a sphere with radius 3,960 miles. Use a calculator and round your answer to the nearest mile.

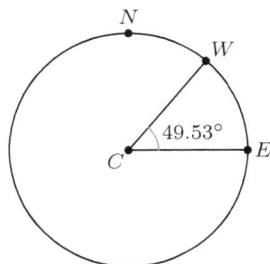

Set 117: Reference Angles

1. Find the measure of the reference angle for the angle of given measure.

A. $2\pi/3$

K. $7\pi/4$

B. $7\pi/6$

L. $3\pi/2$

C. $19\pi/11$

M. $16\pi/7$

D. $5\pi/9$

N. $7\pi/5$

E. $-5\pi/4$

O. $-11\pi/6$

F. $-\pi/3$

P. $-3\pi/4$

G. π

Q. $9\pi/8$

H. $-3\pi/5$

R. $-4\pi/11$

I. $15\pi/4$

S. $35\pi/6$

J. $-34\pi/7$

T. $-31\pi/9$

Set 118: Trigonometric Functions

1. Suppose angles of measures α, β, and γ are drawn in standard position. The terminal side of the angle of measure:
   - α intersects the unit circle at the point $(-4/5, -3/5)$;
   - β intersects the unit circle at the point $(12/13, -5/13)$;
   - γ intersects the unit circle at the point $(0, 1)$.

   Evaluate each expression.

A. $\sin \alpha$

B. $\sec \alpha$

C. $\cot \alpha$

D. $\cos \beta$

E. $\csc \beta$

F. $\tan \beta$

G. $\sec \gamma$

H. $\cos \gamma$

I. $\cot \gamma$

J. $\tan \alpha$

K. $\csc \alpha$

L. $\cos \alpha$

M. $\cot \beta$

N. $\sec \beta$

O. $\sin \beta$

P. $\tan \gamma$

Q. $\csc \gamma$

R. $\sin \gamma$

Set 119: Cosine and Sine of Measures of Acute Angles

1. Evaluate each expression.

A. $\cos \pi/3$ N. $\tan \pi/6$

B. $\csc \pi/2$ O. $\sec \pi/4$

C. $\cot \pi/6$ P. $\sin \pi/2$

D. $\sec 0$ Q. $\cot \pi/3$

E. $\tan \pi/4$ R. $\cos 0$

F. $\sin \pi/4$ S. $\csc \pi/6$

G. $\tan \pi/3$ T. $\cos \pi/6$

H. $\sin 0$ U. $\cot \pi/2$

I. $\cos \pi/2$ V. $\sin \pi/3$

J. $\sec \pi/6$ W. $\csc \pi/4$

K. $\csc \pi/3$ X. $\tan 0$

L. $\cot \pi/4$ Y. $\sec \pi/3$

M. $\sin \pi/6$ Z. $\csc 0$

Set 120: Evaluating Trigonometric Functions

1. Evaluate each expression.

A. $\cos 5\pi/4$

B. $\sin 11\pi/6$

C. $\sin 3\pi/2$

D. $\csc 4\pi/3$

E. $\tan 7\pi/4$

F. $\cos 5\pi/6$

G. $\cos(-5\pi/3)$

H. $\sin(-3\pi/4)$

I. $\cos(-7\pi/2)$

J. $\sec(-4\pi)$

K. $\cot 17\pi/6$

L. $\sin(-25\pi/4)$

M. $\csc 15\pi$

N. $\sin 2\pi/3$

O. $\cos \pi$

P. $\cos 7\pi/6$

Q. $\sec 3\pi/4$

R. $\cot 5\pi/3$

S. $\sec 9\pi/2$

T. $\sin(-7\pi/4)$

U. $\cos(-\pi/6)$

V. $\sin(-6\pi)$

W. $\csc(-5\pi/2)$

X. $\tan 22\pi/3$

Y. $\cos(-25\pi/6)$

Z. $\sec 98\pi$

Set 121: Graphs of the Cosine and Sine Functions

1. State the period p and amplitude a of each function and draw one cycle of its graph.

A. $h(x) = \cos(x - 3\pi/2)$

F. $g(x) = -3 + \sin x$

B. $f(x) = 5 \sin x$

G. $h(x) = (\cos x)/2$

C. $g(x) = \cos(4x/3)$

H. $f(x) = -\sin 6x$

D. $h(x) = (\sin 2x)/8$

I. $g(x) = -3 \cos(x/4)$

E. $f(x) = -2 \cos(\pi(x+1)) + 5$

J. $h(x) = 8 + 4(\sin(x - \pi))/3$

2. The curves below are the graphs of a function f defined by

$$f(x) = A \cdot \cos(B(x - C)) + D$$

for some constants $A, B, C, D \in \mathbf{R}$. For each curve:
- determine the period p and amplitude a of f,
- determine the values of A and B, assuming both are positive,
- determine the values of C and D, and
- write the equation that defines f.

A.

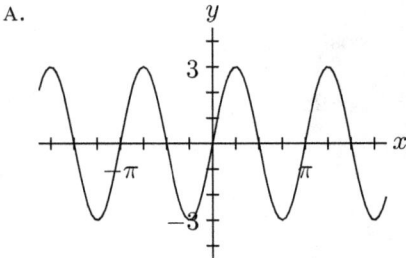

C.

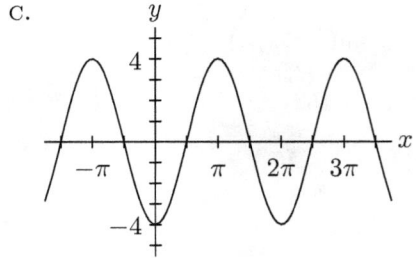

B.

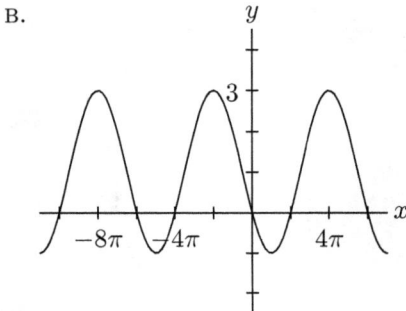

D.

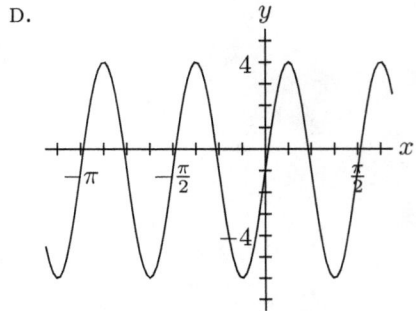

3. The graph of $y = \sin x$ is shown below. Identify the coordinates of points A through E.

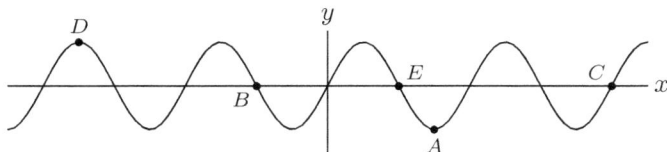

Set 122: Graphs of Other Trigonometric Functions

1. State the period p of each function and draw one cycle of its graph.

A. $g(x) = (\cot x)/2$

B. $h(x) = 4 + \csc x$

C. $f(x) = (\tan(\pi - x))/4$

D. $g(x) = 2\sec(x - \pi/2) - 1$

E. $h(x) = \cot(3x + 3\pi) + 2$

F. $f(x) = -\csc(2(x - 3\pi/4)) - 3$

G. $h(x) = \sec(x + \pi/4)$

H. $f(x) = \tan 3x$

I. $g(x) = -\cot(x - 5\pi)$

J. $h(x) = -\frac{1}{3}\csc(-2x) - 2$

K. $f(x) = 2\tan(\frac{1}{2}(x - \pi/4))$

L. $g(x) = \frac{1}{2}\sec(x + \pi) + 1$

Set 123: Trigonometry of Right Triangles

1. Use the given figure to evaluate the six basic trigonometric functions at θ.

A.

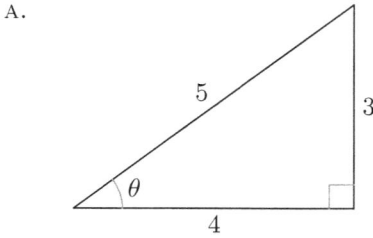

C.

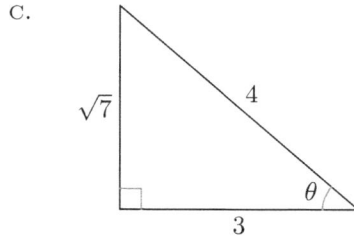

B.

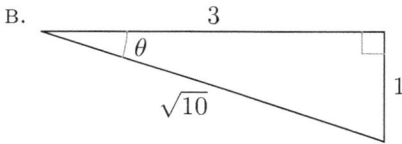

D.

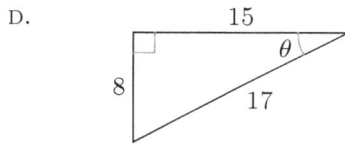

2. Determine the unknown lengths of the sides of each triangle.

A.

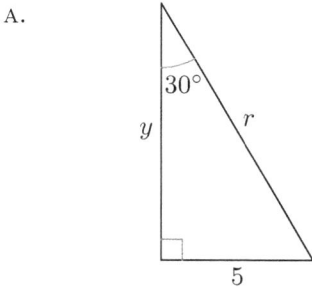

C.

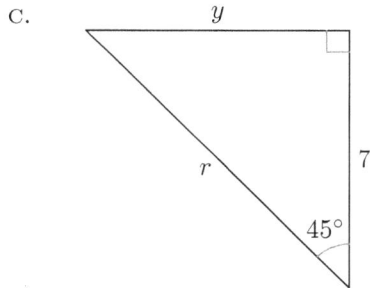

B.

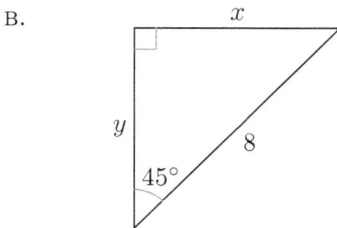

D.

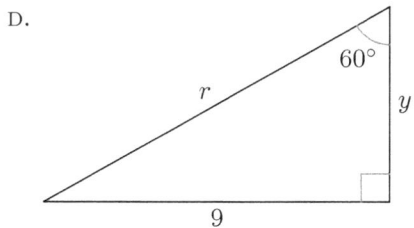

3. Solve each word problem. Use a calculator and round your answers to the nearest tenth of a unit.

A. A cat sits at the top of a telephone pole. The pole is thirty feet high. The angle of elevation from an observer on the ground to the cat is $77°$. How far is the observer standing from the base of the pole?

B. A boy flies a kite at the beach. If he has let out eighty yards of string and the angle of elevation between the boy and the kite is 25°, then how high is the kite? Assume the boy has pulled the string taut.

C. A seagull flying forty feet above the ground sees a woman sunbathing on the beach. The angle of depression from the seagull to the woman is 52°. What is the distance between the seagull and the woman?

D. A woman stands 35 feet from the goal line on a football field. Her son has placed a pin at the point on the goal line nearest her. The woman then errantly bowls a ball toward the pin in a direction that deviates from true aim by 10°. At the time the ball crosses the goal line, how far is it from the pin?

E. A ladder leans against a wall and forms an angle of 67° with the floor. If the ladder is ten feet long, then how far from the wall does the base of the ladder rest?

F. A damsel stands at the top of a burning tower 92 feet above the ground, but will soon slide down a wooden chute to safety on the ground below. If the chute forms an angle of eleven degrees with the ground, then how long is the chute?

4. Use the given information to evaluate the six basic trigonometric functions at θ. Assume that $0° < \theta < 90°$.

A. $\sin \theta = 4/5$ C. $\tan \theta = 12/5$ E. $\cot \theta = 1/8$

B. $\sec \theta = 3/2$ D. $\csc \theta = 6$ F. $\cos \theta = 2/9$

5. The figure at right shows a regular hexagon inscribed in a circle of radius 14 and a right triangle with one vertex at the center of the circle. Given this figure, determine:
   - the measures (in degrees) of the acute angles of the right triangle,
   - the lengths of the legs of the right triangle, and
   - the length of each side of the hexagon.

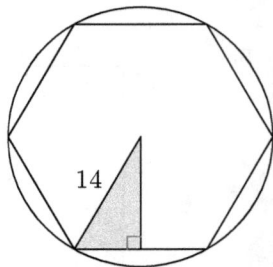

6. Two ducks sit on level ground near a flagpole. Duck B sits c feet closer to the pole than duck A. The angles of elevation from ducks A and B to the top of the flagpole are α and β, respectively. Show that the height of the flagpole is $(c \tan \alpha \tan \beta)/(\tan \beta - \tan \alpha)$ feet.

7. A circle of radius 5 is inscribed within an equilateral triangle (see figure at right). Determine the perimeter of the triangle.

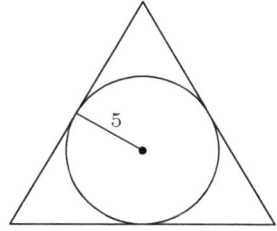

Set 124: Trigonometric Identities

1. Prove each identity.

A. $\sin\theta + \cos\theta\cot\theta = \csc\theta$

H. $\sin^2\theta - \cos^2\theta = 1 - 2\cos^2\theta$

B. $\cot\theta\cos\theta = \csc\theta - \sin\theta$

I. $\tan\theta\csc\theta = \sec\theta$

C. $\cos^4\theta - \sin^4\theta = \cos^2\theta - \sin^2\theta$

J. $(1 - \cos\theta)(\csc\theta + \cot\theta) = \sin\theta$

D. $(\cos\theta)(\sec\theta - \cos\theta) = \sin^2\theta$

K. $\cos^3\theta = \cos\theta - \cos\theta\sin^2\theta$

E. $\dfrac{1}{1 + \cos\theta} = \csc^2\theta - \cot\theta\csc\theta$

L. $\dfrac{\tan\theta}{\sec\theta} = \sin\theta$

F. $\dfrac{1 - \sin\theta}{1 + \sin\theta} = (\sec\theta - \tan\theta)^2$

M. $\dfrac{\sin\theta}{1 + \cos\theta} + \dfrac{1 + \cos\theta}{\sin\theta} = 2\csc\theta$

G. $\dfrac{\sin\theta\sec\theta}{\tan\theta + \cot\theta} = \sin^2\theta$

N. $\dfrac{\cos\theta + \sin\theta}{\sec\theta + \csc\theta} = \cos\theta\sin\theta$

2. Simplify each expression.

A. $\dfrac{\sec\theta}{\tan\theta}$

D. $\dfrac{\sin\theta}{\cot\theta} - \dfrac{\tan\theta}{\csc\theta}$

B. $\cos\theta\sin\theta\csc\theta\sec\theta$

E. $(\cot\theta)/(\cos\theta\csc\theta)$

C. $\dfrac{\sin\theta}{\csc\theta} + \dfrac{\cos\theta}{\sec\theta}$

F. $\dfrac{\cos\theta}{\tan\theta} + \dfrac{\tan\theta}{\sec\theta} - \csc\theta$

3. Determine each value requested in two ways:
- using a trigonometric identity and
- using a right triangle and a reference angle.

A. Evaluate $\sin\theta$ where $\cot\theta = -4/3$ and $3\pi/2 < \theta < 2\pi$.

B. Evaluate $\tan\theta$ where $\cos\theta = -15/17$ and $\pi < \theta < 3\pi/2$.

C. Evaluate $\cos\theta$ where $\sin\theta = 2/5$ and $\pi/2 < \theta < \pi$.

D. Evaluate $\cot\theta$ where $\csc\theta = 13/5$ and $4\pi < \theta < 9\pi/2$.

E. Evaluate $\sec\theta$ where $\tan\theta = 4$ and $3\pi < \theta < 7\pi/2$.

F. Evaluate $\csc\theta$ where $\sec\theta = -\sqrt{5}$ and $\pi/2 < \theta < \pi$.

Set 125: Addition and Subtraction Formulas

1. Prove each identity.

A. $\sin(\theta + \pi) = -\sin\theta$ C. $\cos(\pi - \theta) = -\cos\theta$

B. $\tan(\pi/2 - \theta) = \cot\theta$ D. $\csc(\theta - \pi/2) = -\sec\theta$

E. $\sin(\alpha + \beta) + \sin(\alpha - \beta) = 2\sin\alpha\cos\beta$

F. $\cos(\alpha - \beta)\cos(\alpha + \beta) = \cos^2\beta - \sin^2\alpha$

2. Simplify each expression.

A. $\sin(7\pi/12)\cos(5\pi/12) - \cos(7\pi/12)\sin(5\pi/12)$

B. $\cos(5\pi/8)\cos(3\pi/8) + \sin(5\pi/8)\cos(3\pi/8)$

C. $[\tan(15\pi/7) + \tan(6\pi/7)]/[1 - \tan(15\pi/7)\tan(6\pi/7)]$

D. $\sin 3\theta \sin 8\theta + \cos 3\theta \cos 8\theta$

E. $\sin(\theta + \pi/4) - \cos(\theta - \pi/4)$ F. $\cos(\theta - \pi/6) - \sin(\theta + \pi/3)$

3. Use an addition formula to determine the exact value of each expression.

A. $\cos(5\pi/12)$ C. $\sin(13\pi/12)$

B. $\sin 165°$ D. $\cos(-15°)$

4. Suppose $\sin\alpha = \frac{8}{17}$ and $\cos\beta = -\frac{4}{5}$ where $\frac{\pi}{2} < \alpha < \pi$ and $\pi < \beta < \frac{3\pi}{2}$. Evaluate $\cos(\alpha - \beta)$ and $\sin(\alpha + \beta)$.

5. Suppose $\cos\alpha = -\frac{5}{13}$ and $\sin\beta = -\frac{2}{3}$ where $\frac{\pi}{2} < \alpha < \pi$ and $\frac{3\pi}{2} < \beta < 2\pi$. Evaluate $\cos(\alpha + \beta)$ and $\sin(\alpha - \beta)$.

6. Suppose α, β, and γ represent the measures of the angles of a triangle. Show that $\sin(\alpha + \beta) = \sin\gamma$.

7. Rewrite each expression in terms of sine only.

A. $7\sin x + 7\cos x$ C. $-4\sin x - 4\sqrt{3}\cos x$

B. $-3\sqrt{3}\sin x + 3\cos x$ D. $-2\cos x + 2\sin x$

Set 126: Double-Angle and Half-Angle Formulas

1. Prove each identity.

A. $\cos 2\theta = \sin 2\theta \cot \theta - 1$

F. $\sin 2\theta = 2\tan\theta - 2\tan\theta \sin^2\theta$

B. $\dfrac{2\tan\theta}{1+\tan^2\theta} = \sin 2\theta$

G. $\dfrac{\cot\theta}{\cot\theta - \sin 2\theta} = \sec 2\theta$

C. $\dfrac{\cos 2\theta}{\sin\theta} + \dfrac{\sin 2\theta}{\cos\theta} = \csc\theta$

H. $\dfrac{1+\tan^2\theta}{1-\tan^2\theta} = \sec 2\theta$

D. $\cot\theta - \tan\theta = 2\cot 2\theta$

I. $(\cos\theta + \sin\theta)^2 = 1 + \sin 2\theta$

E. $\cot 2\theta + \tan\theta = \csc 2\theta$

J. $2\csc 2\theta = \cot\theta + \tan\theta$

2. Use the figure shown below to evaluate each expression.

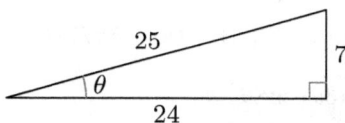

A. $\cos 2\theta$

D. $\cos(\theta/2)$

B. $\sin 2\theta$

E. $\sin(\theta/2)$

C. $\tan 2\theta$

F. $\tan(\theta/2)$

3. Use a half-angle formula to determine the exact value of each expression.

A. $\sin(3\pi/8)$

D. $\cos(7\pi/12)$

B. $\tan 15°$

E. $\sin 165°$

C. $\cos(-75°)$

F. $\tan(-\pi/8)$

4. Suppose $\cos\alpha = -15/17$ where $\pi/2 < \alpha < \pi$.
Evaluate $\cos(\alpha/2)$, $\sin(\alpha/2)$, and $\tan(\alpha/2)$.

5. Suppose $\sin\beta = -12/13$ where $3\pi/2 < \alpha < 2\pi$.
Evaluate $\cos 2\beta$, $\sin 2\beta$, and $\tan 2\beta$.

6. Suppose $\sin 2\theta = c$ and $0 \le \theta \le \pi/2$. Write $\cos\theta + \sin\theta$ in terms of c.

7. Suppose $\cos\theta + \sin\theta = 7/17$ and $\pi/2 \le \theta \le \pi$. Evaluate $\tan\theta$.

8. Write each expression in a form that contains no exponents greater than 1.

A. $\sin^4\theta$

C. $\cos^5\theta$

B. $\cos^2\theta\sin^3\theta$

D. $\tan^6\theta$

9. The figure at right shows a regular octagon inscribed in a circle of radius 6 and a right triangle with one vertex at the center of the circle. Given this figure, determine:
  - the measures (in degrees) of the acute angles of the right triangle,
  - the lengths of the legs of the right triangle, and
  - the length of each side of the octagon.

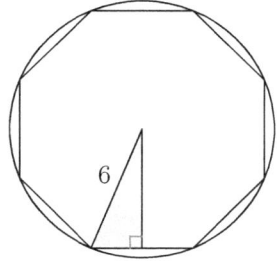

10. Determine each value.

A. the sine of the angle formed by lines $y = (1/2)x$ and $y = 2x$ for $x \geq 0$ [Hint: Consider the triangle with vertices $(0,0)$, $(2,1)$, and $(1,2)$.]

B. the cosine of the angle formed by lines $y = (1/3)x$ and $y = 3x$ for $x \geq 0$

Set 127: Product-to-Sum and Sum-to-Product Formulas

1. Prove each identity.

A. $\dfrac{\sin 6x - \sin 4x}{\cos 6x + \cos 4x} = \tan x$

B. $\dfrac{\sin 7\theta + \sin \theta}{\sin 7\theta - \sin \theta} = \tan 4\theta \cot 3\theta$

C. $\cos(x + y)\sin(x - y) = \sin x \cos x - \sin y \cos y$

D. $\cos n \sin(m + n) - \sin n \cos(m + n) = \sin m$

2. Convert each product to a sum or difference and simplify.

A. $\cos(5\pi/8)\cos(11\pi/8)$

E. $\cos 75° \cos 105°$

B. $\cos 255° \sin 15°$

F. $\sin(\pi/12)\sin(-7\pi/12)$

C. $\sin 5\alpha \sin 2\alpha$

G. $\sin 2t \cos 4t$

D. $\cos(\theta + \pi/4)\cos(\theta - \pi/4)$

H. $\cos(\theta - 5\pi/3)\sin(\theta + 5\pi/3)$

3. Convert each sum or difference to a product and simplify.

A. $\cos(5\pi/12) - \cos(\pi/12)$

D. $\sin 15° + \sin 255°$

B. $\sin 112.5° - \sin(-292.5°)$

E. $\cos(3\pi/8) + \cos(9\pi/8)$

C. $\cos(\theta - \pi) + \cos(\theta + \pi)$

F. $\sin(\theta + 5\pi/4) - \sin(\theta - 15\pi/4)$

4. Suppose $a, b, x \in \mathbf{R}$ with $a - b = \pi/2$. Show that $\sin(x + b) + \cos(x + a) = 0$.

Set 128: Inverse Trigonometric Functions

1. Evaluate each expression.

A. $\arccos(\sqrt{3}/2)$

N. $\arcsin(\sqrt{2}/2)$

B. $\sin^{-1}(-1/2)$

O. $\cos^{-1}0$

C. $\cos^{-1}(-\sqrt{2}/2)$

P. $\sin^{-1}(-\sqrt{3}/2)$

D. $\arcsin(-1)$

Q. $\arccos(1/2)$

E. $\arctan 0$

R. $\tan^{-1}(-\sqrt{3}/3)$

F. $\cot^{-1}(\sqrt{3})$

S. $\cot^{-1}1$

G. $\sec^{-1}(-1)$

T. $\csc^{-1}(-\sqrt{2})$

H. $\csc^{-1}2$

U. $\sec^{-1}(2\sqrt{3}/3)$

I. $\sin(\arcsin 0.2)$

V. $\tan(\tan^{-1}8)$

J. $\arccos(\cos 5\pi)$

W. $\arcsin(\sin(-\pi/8))$

K. $\cot^{-1}(\cot 7\pi/4)$

X. $\csc^{-1}(\csc(-13\pi/6))$

L. $\sec^{-1}(\sec(-5\pi/2))$

Y. $\arctan(\tan 9\pi)$

M. $\cos(\sin^{-1}(1/2) - \tan^{-1}1)$

Z. $\cot(\cos^{-1}0 + \sec^{-1}(-2))$

2. Use a right triangle to evaluate each expression.

A. $\sin(\cot^{-1}(15/8))$

C. $\sec(\tan^{-1}(2/5))$

B. $\csc(\cos^{-1}(4/7))$

D. $\cos(\csc^{-1}3)$

3. Write each expression in a form that contains no trigonometric functions.

A. $\cos(\arcsin(x/2))$ where $-2 < x < 0$

B. $\cot(\csc^{-1}(x/7))$ where $x > 7$

C. $\csc(\tan^{-1}x)$ where $x > 0$

D. $\sin(\sec^{-1}(3/x))$ where $-3 < x < 0$

4. Draw the graph of each function.

A. $f(x) = \pi/6 + \sin^{-1} x$ D. $g(x) = 2\cos^{-1} x$

B. $h(x) = \tan^{-1}(5x)$ E. $f(x) = \csc^{-1}(x - 1)$

C. $g(x) = 3(\sec^{-1} x) - \pi$ F. $h(x) = \cot^{-1}(3 + x/2)$

Set 129: Trigonometric Equations

1. Solve each equation. Leave your solutions in units of radians.

A. $\sin\theta = \sqrt{3}/2$

B. $\cos x = -1/2$

C. $\csc^2 t + 3\csc t + 2 = 0$

D. $\sec^2\alpha - \sec\alpha = 2$

E. $\tan\beta = -1$

F. $2\sin^2 x - 1 = 0$

G. $\sin^2 x + 4\cos x = 4$

H. $\cos\alpha = \sqrt{2}/2$

I. $\sin t = 0$

J. $\csc\theta = -2$

K. $\cos x \cot x = \cos x$

L. $4\tan^2\theta - 12 = 0$

M. $7\cot^2 x - 6\csc^2 x + 6 = 0$

N. $4\cos^2 x = 3$

2. Solve each equation in units of degrees and then determine the solutions of the equation that lie in the interval $[0°, 360°)$.

A. $\cos 2\theta = -\sqrt{3}/2$

B. $\tan(t/2) = -\sqrt{3}$

C. $\csc 4x = -2$

D. $\sin(x/3) = 1/2$

E. $\sec 4m = -1$

F. $\cot 3\theta = \sqrt{3}/3$

3. Find approximations for the solutions of each equation that lie within the interval $[0, 2\pi)$. Use a calculator and round your answers to four decimal places.

A. $\sin\theta = -0.62$

B. $\csc t = 3.54$

C. $\tan 2x = 6.11$

D. $\cos x = -0.18$

E. $\cot\theta = -8.33$

F. $\sec t = 1.46$

4. Find approximations for the solutions of each equation that lie within the interval $[0°, 360°)$. Use a calculator and round your answers to two decimal places.

A. $\cos x = 0.91$

B. $\csc 2\theta = -2.05$

C. $\cot t = 0.74$

D. $\sin t = -0.36$

E. $\tan 4x = -11.68$

F. $\sec 3\theta = -4.49$

5. Find an approximation for the solution of each statement that lies within the interval $[0°, 360°)$. Use a calculator and round each of your answers to two decimal places.

A. $\cos x = -\dfrac{5}{13}$ AND $\sin x = \dfrac{12}{13}$ C. $\cos x = -\dfrac{15}{17}$ AND $\sin x = -\dfrac{8}{17}$

B. $\cos x = \dfrac{\sqrt{11}}{4}$ AND $\sin x = \dfrac{\sqrt{5}}{4}$ D. $\cos x = \dfrac{\sqrt{10}}{5}$ AND $\sin x = -\dfrac{\sqrt{15}}{5}$

6. Rewrite each expression in terms of sine only. Use a calculator where necessary.

A. $-3 \sin x + 5 \cos x$ C. $9 \sin x + 6 \cos x$

B. $\sqrt{5} \sin x - \sqrt{31} \cos x$ D. $-\sqrt{7} \cos x - \sqrt{2} \sin x$

Set 130: Area of a Triangle

1. Determine the area of each triangle below. Use a calculator and round your solutions to two decimal places.

A.

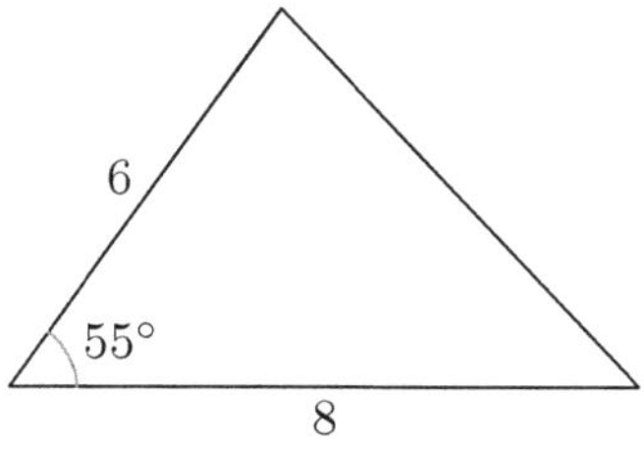

B.

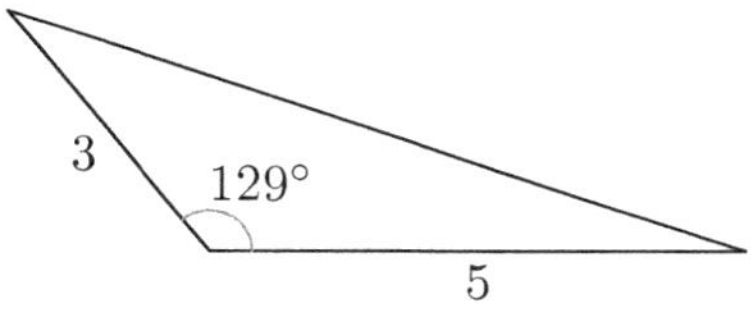

C.

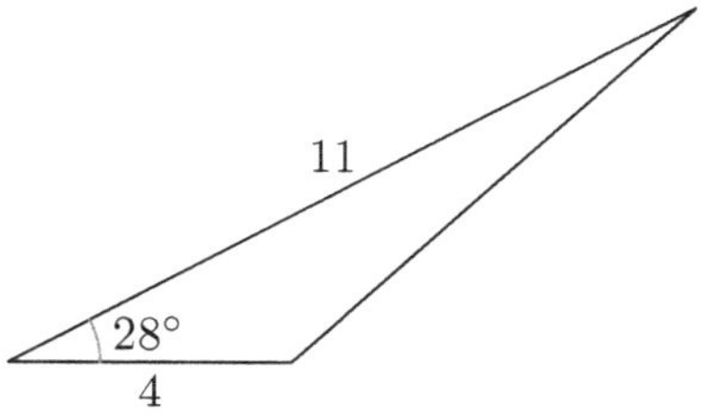

D.

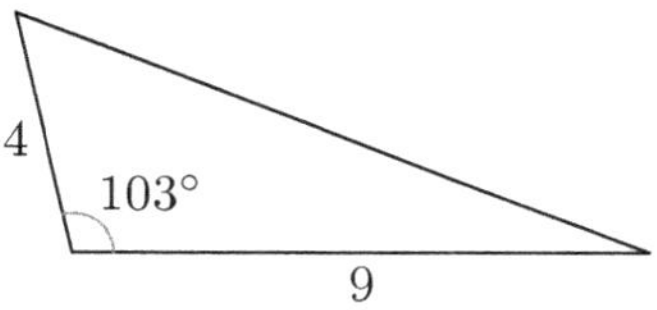

2. Determine the exact area of the region shaded in each figure below. Assume that the polygons are regular and that the radius of the circles is 1.

A.

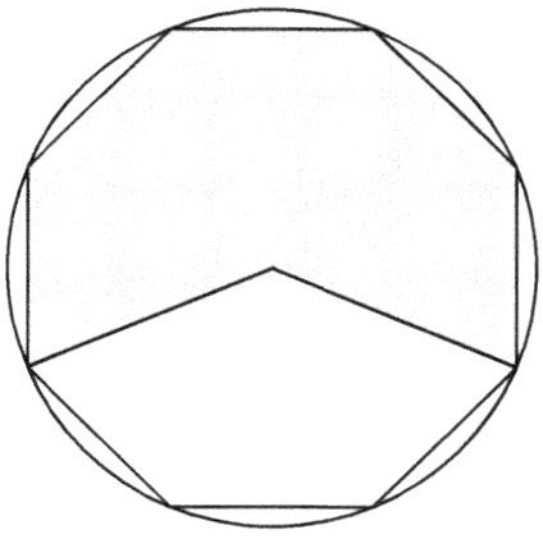

B.

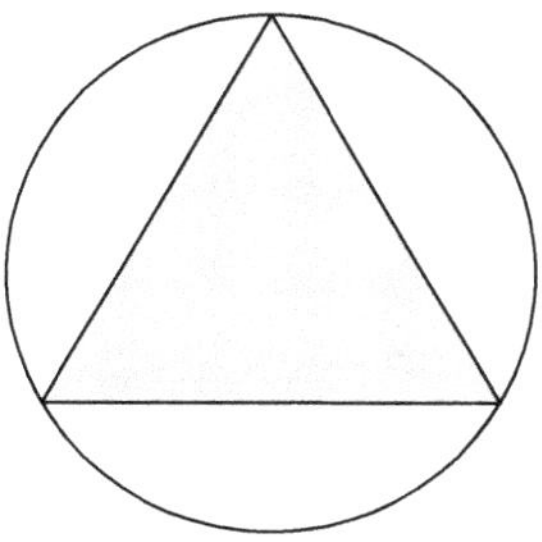

C.

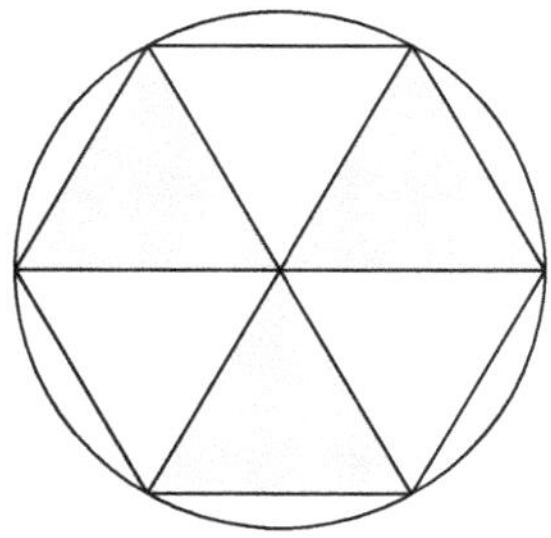

D.

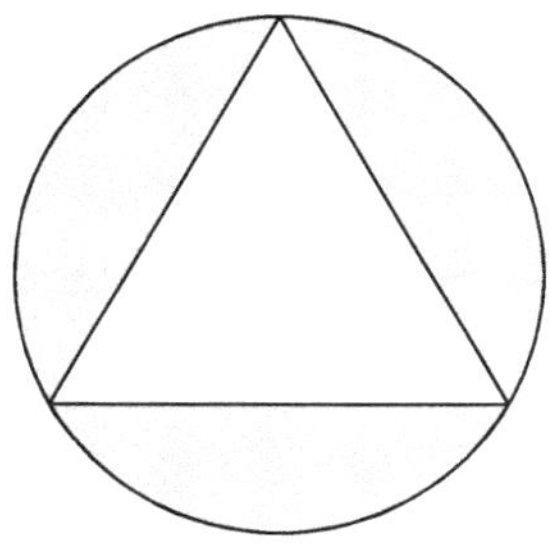

Set 131: Area of a Circular Sector

1. Determine the area of the circular sector bounded by the sides of the angle of measure θ and the circle of radius r centered at the vertex of the angle.

   A. $\theta = \pi/3$, $r = 5$ B. $\theta = 7\pi/6$, $r = 4$

2. Determine the area of the region shaded in the figure below. Assume that the quadrilateral is a square and that the radius of the circle is 1.

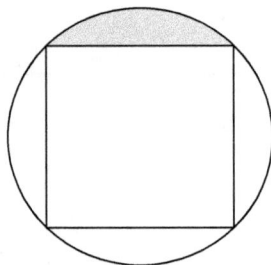

3. Determine the volume of a piece of pie cut from a circular pan if the pan has a diameter of 12 inches and a depth of 2 inches and the central angle of the piece of pie is 15°. Give both the exact value and an approximation rounded to two decimal places.

Set 132: The Law of Sines

1. In the following exercises, assume that a, b, and c represent the lengths of the sides of a triangle that lie opposite the angles of the triangle of measures α, β, and γ, respectively. Use a calculator and round your solutions to two decimal places.

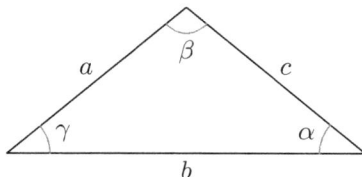

A. Determine a where $\alpha = 44°$, $\beta = 81°$, and $b = 10$.

B. Determine b where $\beta = 53°$, $\gamma = 19°$, and $c = 7$.

C. Determine c where $\alpha = 125°$, $\beta = 22°$, and $a = 9$.

D. Determine α where $a = 5$, $b = 12$, and $\beta = 96°$.

E. Determine β where $a = 4$, $c = 3$, $\gamma = 40°$, and α is obtuse.

F. Determine all possible values of γ where $a = 5$, $c = 8$, and $\alpha = 34°$ and then draw a triangle with sides of lengths a and c and angles of measures α and γ for each possible value of γ.

2. A lighthouse sits on a cliff above the sea. The lighthouse is 35 yards tall. The angles of depression from the top and the bottom of the lighthouse to a tennis ball at sea are 19° and 16°, respectively. Determine the distance from the bottom of the lighthouse to the ball.

Set 133: The Law of Cosines

1. In the following exercises, assume that a, b, and c represent the lengths of the sides of a triangle that lie opposite the angles of the triangle of measures α, β, and γ, respectively. Use a calculator and round your solutions to two decimal places.

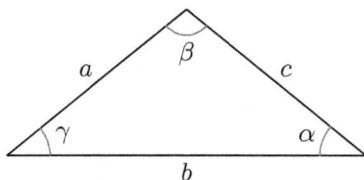

A. Determine a where $b = 4.6$, $c = 5.9$, and $\alpha = 31°$.

B. Determine b where $a = 9.6$, $c = 12.8$, and $\beta = 154°$.

C. Determine c where $a = 10.3$, $b = 11.5$, and $\beta = 49°$.

D. Determine α where $a = 7.6$, $b = 8.9$, and $c = 9.4$

E. Determine β where $a = 7.1$, $b = 13.7$, and $c = 6.8$.

F. Determine α, β, and γ where $a = 6$, $b = 2$, and $c = 7$ and then draw a triangle with sides of lengths a, b, and c and angles of measures α, β, and γ.

2. Consider $\triangle PQR$ where point M is the midpoint of QR. Determine $d(Q, R)$ if $d(P, Q) = 7$, $d(P, R) = 9$, and $d(P, M) = 4$.

3. A point lies inside an equilateral triangle at distances of 3, 4, and 5 units from its vertices. Determine the area of the triangle.

Set 134: Heron's Formula

1. Given the lengths a, b, and c of the sides of a triangle:
   - find the area of the triangle using Heron's Formula,
   - find the measure θ of the angle opposite the side of length c, and
   - find the area of the triangle using the formula $A = (1/2)ab \sin \theta$.

   Use a calculator and round your answers to two decimal places.

   A. $a = 6$, $b = 9$, $c = 7$ C. $a = 5$, $b = 8$, $c = 11$

   B. $a = 10$, $b = 7$, $c = 14$ D. $a = 12$, $b = 11$, $c = 6$

2. Suppose that a, b, and c represent the lengths of the sides of a triangle and that $s = (a + b + c)/2$. Prove that $s > c$.

3. Suppose $\triangle ABC$ has sides of lengths 4, 7, and 7 and $\triangle DEF$ has sides of lengths 4, 7, and 9. Which triangle has the greater area?

Set 135: Conic Sections

1. Identify the conic sections indicated in the figure below.

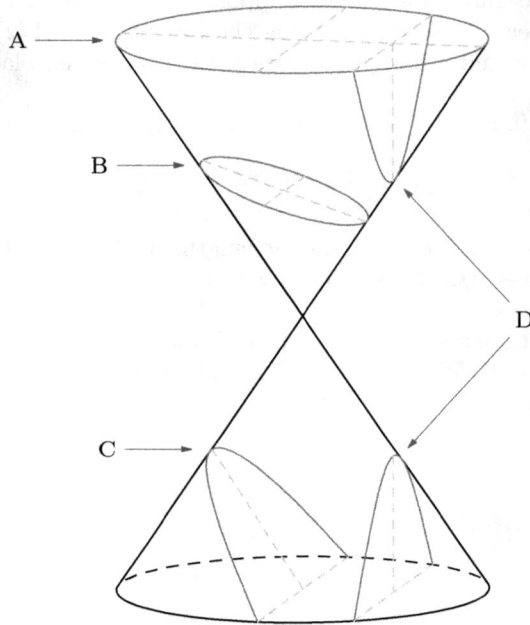

Set 136: Parabolas

1. Determine the equation of the parabola with vertex $(0,0)$ and the given characteristics.

A. focus has coordinates $(4,0)$ E. focus has coordinates $(0,5)$

B. focus has coordinates $(0,-1)$ F. focus has coordinates $(-6,0)$

C. directrix has equation $y = 2$ G. directrix has equation $x = 7$

D. directrix has equation $x = -3$ H. directrix has equation $y = 8$

I. the focus lies on the y-axis;
the focus lies above the x-axis;
the distance from the focus to the directrix is 10

J. the focus lies on the x-axis;
the focus lies to the left of the y-axis;
the distance from the focus to the directrix is 4

K. the focus lies on the x-axis;
the point $(3,-6)$ lies on the parabola

L. the focus lies on the y-axis;
the point $(12,18)$ lies on the parabola

M. the focus lies on the x-axis;
the focus lies to the right of the y-axis;
focal diameter $= 8$

N. the focus lies on the y-axis;
the focus lies below the x-axis;
focal diameter $= 5$

O. the directrix has a negative y-intercept;
focal diameter $= 11$

P. the directrix has a positive x-intercept;
focal diameter $= 10$

2. Identify the focus, directrix, vertex, and focal diameter of the parabola with the given equation.

A. $x^2 = -6y$

D. $y^2 = 9x$

B. $y^2 - 12y - 2x = 0$

E. $x^2 + 2x - 4y + 13 = 0$

C. $x^2 - 3x = 5y + 8$

F. $y^2 + 5y = -8x - 7$

3. Determine the equation and draw a rough sketch of the parabola with the given characteristics. Include the vertex, focus, directrix, and latus rectum in your sketch.

A. focus has coordinates $(0, 4)$;
  vertex has coordinates $(0, 0)$;

D. focus has coordinates $(-3, 0)$;
  directrix has equation $x = 3$

B. focus has coordinates $(4, 5)$;
  directrix has equation $x = 2$

E. focus has coordinates $(-8, -1)$;
  vertex has coordinates $(-8, 2)$;

C. directrix has equation $y = -2$;
  vertex has coordinates $(8, 5)$

F. directrix has equation $x = -3$;
  vertex has coordinates $(-6, 1)$

Set 137: Ellipses

1. Rewrite each equation in the form $(x - h)^2/m^2 + (y - k)^2/n^2 = 1$.

A. $9x^2 + 4y^2 = 36$ D. $50x^2 + 32y^2 = 2$

B. $4x^2 + 8x + 25y^2 - 100y = -103$ E. $x^2 - 2x + 16y^2 + 80y + 37 = 0$

C. $3x^2 - 24x + 7y^2 - 14y = -34$ F. $5x^2 + 15x + 2y^2 + 20y + 31 = 0$

2. Determine the equation of the ellipse with the given characteristics.

A. foci: $(1, 3)$ and $(1, 9)$;   B. foci: $(-18, 5)$ and $(12, 5)$;
   vertices: $(1, 1)$ and $(1, 11)$ vertices: $(-20, 5)$ and $(14, 5)$

C. the endpoints of the major axis are $(-17, -4)$ and $(-3, -4)$;
   the endpoints of the minor axis are $(-10, 1)$ and $(-10, -9)$

D. the endpoints of the major axis are $(2, -2)$ and $(2, 10)$;
   the endpoints of the minor axis are $(-1, 6)$ and $(5, 6)$

E. the vertices lie on the x-axis;
   the center lies at the origin
   the point $(3, -\sqrt{3}/2)$ lies on the ellipse;
   the length of the major axis is twice the length of the minor axis

F. the vertices lie on the y-axis;
   the center lies at the origin
   the point $(\sqrt{34}/5, 4)$ lies on the ellipse;
   the length of the major axis is five times the length of the minor axis

G. the center lies at the origin;
   the eccentricity is $5/13$;
   one endpoint of the major axis has coordinates $(0, 26)$

H. the center has coordinates $(-1, 4)$;
   the eccentricity is $9/10$;
   one focus has coordinates $(8, 4)$

3. Find the x-coordinates of all points on the graph of $x^2/8 + y^2/50 = 1$ with a y-coordinate of 5.

4. Find the y-coordinates of all points on the graph of $x^2/6 + y^2/16 = 1$ with an x-coordinate of 2.

5. Draw the ellipse with the given characteristics.

A. center: $(0,0)$;
   foci lie on the x-axis;
   length of major axis $= 12$;
   length of minor axis $= 2$;

D. center: $(0,0)$;
   foci lie on the y-axis;
   length of major axis $= 7$;
   length of minor axis $= 4$;

B. center: $(8,5)$;
   foci lie on a vertical line;
   length of major axis $= 8$;
   length of minor axis $= 3$;

E. center: $(1,-4)$;
   foci lie on a horizontal line;
   length of major axis $= 9$;
   length of minor axis $= 6$;

C. center: $(-7,0)$;
   foci lie on a horizontal line;
   length of major axis $= 4$;
   length of minor axis $= 1$;

F. center: $(-3,-2)$;
   foci lie on a vertical line;
   length of major axis $= 10$;
   length of minor axis $= 5$;

6. Identify the foci, vertices, center, eccentricity, and lengths of the major and minor axes of the ellipse with the given equation and draw its graph.

A. $x^2/4 + y^2/49 = 1$

E. $x^2/81 + y^2/36 = 1$

B. $16x^2 + y^2 = 64$

F. $9x^2 + 25y^2 = 225$

C. $(x-2)^2/25 + (y-7)^2/9 = 1$

G. $(x+5)^2/100 + y^2/16 = 1$

D. $(x+6)^2 + (y+2)^2/121 = 1$

H. $(x-8)^2/2 + (y+3)^2/5 = 1$

Set 138: Hyperbolas

1. Rewrite each equation in either the form $(x - h)^2/a^2 - (y - k)^2/b^2 = 1$ or the form $(y - k)^2/a^2 - (x - h)^2/b^2 = 1$.

A. $y^2 - 4y - x^2 - 18x - 78 = 0$ D. $x^2 + 10x - y^2 + 6y = 0$

B. $x^2 - 5x - y^2 + y - 3 = 0$ E. $8x - x^2 + y^2 - 3y - 26 = 0$

C. $x^2 + 6x - 3y^2 + 12y = 15$ F. $3y^2 - 5y - 2x^2 - 14x = 69/2$

2. Determine the equation of the hyperbola with the given characteristics.

A. foci: $(-3, 2)$ and $(7, 2)$;
   eccentricity: $5/3$

B. foci: $(8, 3)$ and $(8, 15)$;
   eccentricity: 2

C. the vertices lie on the same horizontal line;
   the x-coordinates of the vertices are -5 and -1;
   the asymptotes have equations $5x - 2y = -27$ and $5x + 2y = -3$

D. the vertices lie on the same vertical line;
   the y-coordinates of the vertices are -2 and 4;
   the asymptotes have equations $3x + 4y = 28$ and $3x - 4y = 20$

E. the center lies at $(-9, -4)$;
   the length of the transverse axis is 7;
   the length of the conjugate axis is 5;
   the foci lie on the same vertical line

F. the center lies at $(0, 11)$;
   the length of the transverse axis is $\sqrt{3}$;
   the length of the conjugate axis is 10;
   the foci lie on the same horizontal line

G. the vertices lie at $(-8, -3)$ and $(-8, -11)$;
   the point $(-3, -2)$ lies on the hyperbola

H. the vertices lie at $(-5, 5)$ and $(1, 5)$;
   the point $(3, -11)$ lies on the hyperbola

3. Find the y-coordinates of all points on the graph of $y^2/25 - x^2/2 = 1$ with an x-coordinate of 4.

4. Find the x-coordinates of all points on the graph of $x^2/7 - y^2/5 = 1$ with a y-coordinate of 10.

5. Draw the hyperbola with the given characteristics.

A. center: $(0,0)$;
  foci lie on the x-axis;
  length of transverse axis $= 14$;
  length of conjugate axis $= 2$;

D. center: $(0,0)$;
  foci lie on the y-axis;
  length of transverse axis $= 10$;
  length of conjugate axis $= 18$;

B. center: $(-4,7)$;
  foci lie on a vertical line;
  length of transverse axis $= 11$;
  length of conjugate axis $= 20$;

E. center: $(0,8)$;
  foci lie on a horizontal line;
  length of transverse axis $= 7$;
  length of conjugate axis $= 5$;

C. center: $(6,-1)$;
  foci lie on a horizontal line;
  length of transverse axis $= 1$;
  length of conjugate axis $= 16$;

F. center: $(-14,-5)$;
  foci lie on a vertical line;
  length of transverse axis $= 12$;
  length of conjugate axis $= 4$;

6. Identify the foci, vertices, center, eccentricity, and lengths of the transverse and conjugate axes of the hyperbola with the given equation and draw its graph.

A. $(x+3)^2/9 - (y-10)^2/81 = 1$

E. $(y+4)^2/6 - (x-11)^2 = 1$

B. $x^2 - 8y^2 = 64$

F. $9y^2 - 4x^2 = -36$

C. $25y^2 - 2(x+5)^2 - 4 = 0$

G. $3(x-2)^2 - 5(y-8)^2 = 75$

D. $11(y-7)^2 - 4(x-6)^2 = 1$

H. $-2x^2 + 8(y+1)^2 - 18 = 0$

Set 139: Parametric Equations

1. For each set of parametric equations, determine the point (x, y) on its graph that correponds to the given value of the parameter.

A. $\{x = 4t^2, \ y = -5t^3\}; \ t = 2$

B. $\{x = \tan m, \ y = \csc m\}; \ m = 5\pi/3$

C. $\{x = \sin 3n, \ y = \cos 2n\}; \ n = \pi/6$

D. $\{x = \log b, \ y = b/4\}; \ b = 100$

2. Eliminate the parameter from each set of parametric equations and draw its graph.

A. $\{x = 3t, \ y = t + 4\}$

B. $\{x = t^2 - 8, \ y = t - 5\}$

C. $\{x = t - 6, \ y = t^3\}$

D. $\{x = 10 \sin t, \ y = 10 \cos t\}$

E. $\{x = 9 - t, \ y = -|t| + 1\}$

F. $\{x = e^t, \ y = -e^{-t}\}$

G. $\{x = 5 \cos t + 2, \ y = 11 \sin t - 6\}$

H. $\{x = 6 \sec t - 1, \ y = 2 \tan t + 4\}$
  for $t \notin \{\pi/2 + \pi k \mid k \in \mathbf{Z}\}$

I. $\{x = 4 - t, \ y = t + 9\}$

J. $\{x = t + 2, \ y = t^2 - 4\}$

K. $\{x = |t|, \ y = t + 3\}$

L. $\{x = 5 - \cot t, \ y = -2 + \cot t\}$

M. $\{x = \sin 2t, \ y = 3 \cos 2t\}$

N. $\{x = e^t, \ y = e^{t/2}\}$

O. $\{x = 8 + 3 \cos t, \ y = -4 + 3 \sin t\}$

P. $\{x = 9 \cot t + 5, \ y = 4 \csc t + 2\}$
  for $t \notin \{\pi k \mid k \in \mathbf{Z}\}$

Q. $\{x = \sec t - \tan t, \ y = \sec t + \tan t\}$ for $t \notin \{\pi/2 + \pi k \mid k \in \mathbf{Z}\}$

R. $\{x = \cot t + \csc t, \ y = \cot t - \csc t\}$ for $t \notin \{\pi k \mid k \in \mathbf{Z}\}$

Set 140: Polar Coordinates

1. Plot the points with the given polar coordinates. Use polar graph paper.

A. $(4, 7\pi/4)$
D. $(1, 13\pi/6)$

B. $(3, -\pi)$
E. $(-5, 3\pi/2)$

C. $(2, 4\pi/3)$
F. $(\pi, -5\pi/4)$

2. Find the polar coordinates (r, θ) of the point with the rectangular coordinates (x, y), where r and θ satisfy the conditions $0 \le \theta < 2\pi$ and $r > 0$.

A. $(x, y) = (-8, 8)$
D. $(x, y) = (5, 0)$

B. $(x, y) = (0, -4)$
E. $(x, y) = (1, \sqrt{3})$

C. $(x, y) = (-6, -2\sqrt{3})$
F. $(x, y) = (2, -2\sqrt{3})$

3. Find the rectangular coordinates (x, y) of the point with the polar coordinates (r, θ).

A. $(r, \theta) = (5, \pi/4)$
D. $(r, \theta) = (2, 7\pi/6)$

B. $(r, \theta) = (7, -\pi/2)$
E. $(r, \theta) = (6, 11\pi)$

C. $(r, \theta) = (10, -4\pi/3)$
F. $(r, \theta) = (-4, 3\pi/2)$

4. Identify two other pairs (r, θ) of polar coordinates, one with $r > 0$ and one with $r < 0$, that also provide the location of the point with the given polar coordinates.

A. $(6, 4\pi/3)$
C. $(1, \pi/4)$

B. $(4, -5\pi/6)$
D. $(-11, \pi/2)$

5. Convert the given equation from rectangular form to polar form. Leave your answers in the form $r = f(\theta)$ or $r^2 = f(\theta)$ for some function f.

A. $x^2 + y^2 = 16$
E. $x = 7$

B. $y = 2$
F. $x - 3y = 10$

C. $9x + 4y = -3$
G. $4xy = 20$

D. $y^2 - x^2 = 49$
H. $3x^2 + 2y^2 = 25$

6. Convert the given equation from polar form to rectangular form and describe its graph.

A. $r = 8$

H. $\theta = \pi/4$

B. $\theta = -4\pi/3$

I. $r = -5$

C. $r = 4\cos\theta$

J. $r = 14\sin\theta$

D. $r = -7\csc\theta$

K. $r = 3\sec\theta$

E. $r = 4/\cos(\theta - 7\pi/4)$

L. $r = -8/\sin(\theta + \pi/3)$

F. $r^2 = 30/(2 + 5\sin^2\theta)$

M. $r^2 = 11/(1 - 3\cos^2\theta)$

G. $r = 12\cos\theta - 4\sin\theta$

N. $r = 2\sin\theta + 6\cos\theta$

Set 141: Graphs of Polar Equations

1. Draw the graph of each polar equation. Use polar graph paper.

A. $r = 2$

N. $r = -7$

B. $\theta = 5\pi/6$

O. $\theta = -3\pi/4$

C. $r = 5 \sec \theta$

P. $r = -8 \csc \theta$

D. $r = -2\theta$ for $\theta \geq 0$

Q. $r = \theta/\pi$ for $\theta \geq 0$

E. $r = 8 \sin \theta$

R. $r = -6 \cos \theta$

F. $r = 9/(5 \cos \theta + \sin \theta)$

S. $r = 14/(2 \sin \theta - 7 \cos \theta)$

G. $r = 2 \cos 2\theta$

T. $r = \sin 2\theta$

H. $r = 4 \sin 3\theta$

U. $r = 3 \cos 5\theta$

I. $r = 5 + \cos \theta$

V. $r = 3 - \sin \theta$

J. $r = 3 - 2 \cos \theta$

W. $r = 5 + 4 \sin \theta$

K. $r = 5 + 9 \sin \theta$

X. $r = 3 + 8 \cos \theta$

L. $r^2 = 4 \sin 2\theta$

Y. $r^2 = 9 \cos \theta$

M. $r^2 = 16 \cos 4\theta$

Z. $r^2 = 49 \sin 4\theta$

Set 142: Rotation of Axes

1. For each point with the given rectangular coordinates (x, y) in the system with axes unrotated, find the coordinates (X, Y) of the point in the system with axes rotated through the angle of measure θ.

   A. $(x, y) = (5, -6)$; $\theta = 45°$ B. $(x, y) = (8, 1)$; $\theta = 30°$

2. For each point with the given rectangular coordinates (X, Y) in the system with axes rotated by the angle of measure θ, find the coordinates (x, y) of the point in the system with axes unrotated.

   A. $(X, Y) = (-10, -2)$; $\theta = 60°$ B. $(X, Y) = (-3, 7)$; $\theta = 15°$

3. For each equation of the form $ax^2 + bxy + cy^2 + dx + ey + f = 0$:
   • identify whether the graph of the equation is a parabola, an ellipse, or a hyperbola;
   • determine the values of $\cos 2\theta$, $\cos \theta$, $\sin \theta$, and θ where $0° < \theta < 90°$ and $\cot 2\theta = (a - c)/b$;
   • transform the equation into an equation of the form
   $$AX^2 + CY^2 + DX + EY + F = 0$$
   by substituting $X \cos \theta - Y \sin \theta$ for x and $X \sin \theta + Y \cos \theta$ for y;
   • rewrite the equation in X and Y in the standard form of a parabola, ellipse, or hyperbola as appropriate;
   • draw the graph of the equation on axes rotated through the angle of measure θ.

   A. $xy - 18 = 0$

   B. $x^2 + 2xy + y^2 + x - y - 3 = 0$

   C. $29x^2 - 7xy + 5y^2 - 118 = 0$

   D. $x^2 + 4xy + 4y^2 + (8\sqrt{5})x + (21\sqrt{5})y + 115 = 0$

   E. $33x^2 - 8xy + 18y^2 + (6\sqrt{17})x + (24\sqrt{17})y + 17 = 0$

   F. $x^2 + 6xy - 7y^2 + (6\sqrt{10}/5)x + (82\sqrt{10}/5)y - 14 = 0$

   G. $4x^2 - (2\sqrt{3})xy + 2y^2 + (-8 - 5\sqrt{3})x + (5 - 8\sqrt{3})y - 31 = 0$

   H. $7x^2 + 12xy + 2y^2 + (90\sqrt{13}/13)x + (8\sqrt{13}/13)y - 51 = 0$

   I. $10x^2 - (8\sqrt{5})xy + 8y^2 + (-6 + 36\sqrt{5})x + (-72 - 3\sqrt{5})y + 99 = 0$

Set 143: Complex Numbers

1. For each complex number:
- write the number in the form $a + bi$,
- identify the real part of the number, and
- identify the imaginary part of the number.

A. $9i - 5$

E. $i + 2$

B. $12 - 6i$

F. $1 - 5i$

C. $-4i$

G. $7i$

D. 3

H. -10

2. Convert each of the following to the form bi.

A. $\sqrt{-144}$

D. $\sqrt{-81}$

B. $\sqrt{-18}$

E. $\sqrt{-24}$

C. $\sqrt{-200}$

F. $\sqrt{-72}$

3. Add or multiply the complex numbers as indicated.

A. $(4 + 11i) + (3 + 8i)$

G. $(7 + 2i) + (-1 + 6i)$

B. $(-6 - 5i) + (-9 - 14i)$

H. $(13 - 8i) + 2i$

C. $(1 + 4i) \cdot (7 + 3i)$

I. $(6 - 3i) \cdot (-8 + 5i)$

D. $9i \cdot (2i - 11)$

J. $(-5 - 12i) \cdot (-4i)$

E. $(10 - 4i) \cdot 2$

K. $-8 \cdot (-13 + i)$

F. $(8 + i) \cdot (8 - i)$

L. $(-5 + 4i) \cdot (5 + 4i)$

4. Perform the indicated operations.

A. $\sqrt{-3} \cdot \sqrt{-12}$

E. $\sqrt{-16} \cdot \sqrt{-4}$

B. $\sqrt{-5} \cdot \sqrt{-2}$

F. $\sqrt{-3} \cdot \sqrt{-15}$

C. $4 \cdot \sqrt{-27}$

G. $-7 \cdot \sqrt{-48}$

D. $\sqrt{6} \cdot \sqrt{-14}$

H. $\sqrt{-10} \cdot \sqrt{35}$

5. Simplify each of the following.

A. i^{14} C. i^9

B. i^{23} D. i^{76}

6. Identify the complex conjugate of each complex number.

A. $9 + 2i$ G. $-8 - 5i$

B. $4 - 3i$ H. $-2 + 11i$

C. $7i - 1$ I. $-6i + 10$

D. $-12i$ J. $4i$

E. 8 K. -7

F. 0 L. $-i$

Set 144: Properties of Operations on Complex Numbers

1. For each complex number:
  - state its additive inverse,
  - state its multiplicative inverse, and
  - verify that the product of the number and its multiplicative inverse equals 1.

A. $2 + 4i$ E. $5 - 3i$

B. $8 - i$ F. $9 + 2i$

C. -12 G. 17

D. $7i$ H. $-i$

2. Determine all complex numbers z for which $-z = z^{-1}$.

3. Determine the values of $m, n \in \mathbf{R}$ for which each statement is true.

A. $10 - 24i = 2m + 3ni$ B. $18i - 13 = 5m - 4ni$

4. Prove each statement.

A. $1 \cdot z = z$ for all $z \in \mathbf{C}$

B. $z_1 \cdot z_2 = z_2 \cdot z_1$ for all $z_1, z_2 \in \mathbf{C}$

C. $(z_1 + z_2) + z_3 = z_1 + (z_2 + z_3)$ for all $z_1, z_2, z_3 \in \mathbf{C}$

D. $z_1 \cdot (z_2 + z_3) = z_1 \cdot z_2 + z_1 \cdot z_3$ for all $z_1, z_2, z_3 \in \mathbf{C}$

E. $(\overline{z})^2 = \overline{z^2}$ for all $z \in \mathbf{C}$

5. Suppose $z \in \mathbf{C}$ with $z \neq 0$.
Prove that $z = -\overline{z}$ if and only if z is a pure imaginary number.

Set 145: Subtraction and Division of Complex Numbers

1. Subtract the complex numbers as indicated.

A. $(9 + 5i) - (4 + 7i)$

E. $(6 - 4i) - (-1 - 8i)$

B. $(-11 + 3i) - (4 - 6i)$

F. $(-3 - 12i) - (-5 + 2i)$

C. $(3 - 6i) - 4i$

G. $i - (9 + 3i)$

D. $8 - (17 + 11i)$

H. $(-2 + 8i) - 10$

2. Divide the complex numbers as indicated.
Leave the result in the form $a + bi$.

A. $(8 - 3i) \div (5 + 2i)$

E. $(1 + 9i) \div (7 - i)$

B. $(10 + 4i) \div (1 - 4i)$

F. $(2 - 6i) \div (10 + 8i)$

C. $(13 - 7i) \div 6i$

G. $(11 + 5i) \div i$

D. $(7 + 18i) \div 3$

H. $(6 + 13i) \div (-4)$

Set 146: Complex Solutions of Quadratic Equations

1. Solve each equation.

A. $x^2 - 3x + 4 = 0$ E. $x^2 + 5x + 7 = 0$

B. $2x^2 + 2x + 1 = 0$ F. $3x^2 - 7x + 5 = 0$

C. $x^2 + 9 = 0$ G. $6x^2 = -96$

D. $5x^2 - 10x = -7$ H. $4x^2 = -3 - 6x$

2. Determine the discriminant and the nature of the solutions of each quadratic equation.

A. $x^2 - 5x + 8 = 0$ E. $x^2 - 14x + 49 = 0$

B. $2x^2 - 10x - 9 = 0$ F. $-2x^2 - 4x - 11 = 0$

C. $9x^2 + 30x + 25 = 0$ G. $3x^2 - 7x + 4 = 0$

D. $3x^2 = 11$ H. $5x^2 = -9x - 5$

Set 147: Complex Zeros of Polynomials

1. Find the zeros and multiplicities of the zeros of each polynomial $f(x)$.

A. $f(x) = 3x^5(x+8)^2(x+4)^7(x-2)^6$

B. $f(x) = -5(x+2)^4(x-3)^9(x-4)(x-10)^6$

C. $f(x) = -4(x-8i)^3(x+8i)^3(x+11)^5(x-17)$

D. $f(x) = (x+4-7i)^5(x+4+7i)^5x^{22}(x-3)^6(x-1)^2$

2. Determine the polynomial $f(x)$ with the given characteristics. Assume the polynomial has real coefficients and leave it in expanded form.

A. degree: 3;
zeros: $-7i$, $7i$, 4;
leading coefficient: -5

B. degree: 7;
zeros: $-1+i$, $3i$, 2;
leading coefficient: 1;
the zero 2 has multiplicity 3

C. degree: 4;
zeros: -4, 0, $2i$;
leading coefficient: 3

D. degree: 6;
zeros: $3+i$, -1, 6;
leading coefficient: -2;
zeros -1 and 6 have multiplicity 2

3. Factorize each polynomial completely.

A. $f(x) = 3x^2 + 48$

B. $f(x) = x^2 + 2x + 4$

C. $f(x) = 2x^3 - 54$

D. $f(x) = x^3 - 3x^2 + 25x - 75$

E. $f(x) = x^3 + x^2 - 7x + 65$

F. $f(x) = -2x^2 - 18$

G. $f(x) = 7x^2 - 5x + 1$

H. $f(x) = -3x^3 - 24$

I. $f(x) = x^3 + 7x^2 + 4x + 28$

J. $f(x) = 2x^3 + 23x^2 + 62x - 37$

K. $f(x) = 3x^4 - 7x^3 + 36x^2 + 82x + 36$

L. $f(x) = x^4 + x^3 - 16x^2 - 26x + 40$

Set 148: Graphical Representation of Complex Numbers

1. Construct a portion of the complex plane that will contain the points that correspond to the numbers below and then plot those points in the plane.

A. $z_a = 4 - 3i$

D. $z_d = -5 - 2i$

B. $z_b = -1 + 5i$

E. $z_e = 5 + i$

C. $z_c = 7$

F. $z_f = -4i$

Set 149: Trigonometric Form of Complex Numbers

1. Write each complex number in rectangular form.

A. $12(\cos(\pi/4) + i\sin(\pi/4))$

D. $3(\cos(7\pi/6) + i\sin(7\pi/6))$

B. $5(\cos(5\pi/3) + i\sin(5\pi/3))$

E. $8(\cos(\pi/2) + i\sin(\pi/2))$

C. $7(\cos\pi + i\sin\pi)$

F. $6(\cos(3\pi/4) + i\sin(3\pi/4))$

2. Write each complex number in trigonometric form where the argument lies between 0 and 2π.

A. $-4i$

D. 11

B. $-3/2 + 3i\sqrt{3}/2$

E. $7\sqrt{3} + 7i$

C. $-5\sqrt{2} - 5i\sqrt{2}$

F. $1 - i\sqrt{3}$

3. Determine the product $z_1 z_2$ and the quotient z_1/z_2 of each pair of complex numbers z_1 and z_2. Leave your answers in trigonometric form.

A. $z_1 = 15(\cos(5\pi/6) + i\sin(5\pi/6))$;
$z_2 = 5(\cos(\pi/4) + i\sin(\pi/4))$

B. $z_1 = 2\sqrt{5}(\cos(4\pi/3) + i\sin(4\pi/3))$;
$z_2 = 4\sqrt{5}(\cos(11\pi/6) + i\sin(11\pi/6))$

C. $z_1 = \sqrt{12}(\cos 85° + i\sin 85°)$;
$z_2 = \sqrt{3}(\cos 320° + i\sin 320°)$

D. $z_1 = 18(\cos 190° + i\sin 190°)$;
$z_2 = 2(\cos 110° + i\sin 110°)$

4. Suppose $z_1 = 6 + 6i\sqrt{3}$ and $z_2 = 2\sqrt{2} - 2i\sqrt{2}$. Find the product $z_1 z_2$ and the quotient z_1/z_2 in two ways:
- leaving z_1 and z_2, the product, and the quotient in rectangular form;
- converting z_1 and z_2 to trigonometric form, multiplying and dividing, and converting the product and quotient to rectangular form.

Set 150: Powers and Roots of Complex Numbers

1. Use DeMoivre's Theorem to evaluate each expression. Leave your answer in the same form as the base of the power.

A. $[2(\cos(\pi/4) + i\sin(\pi/4))]^3$

E. $[3(\cos(2\pi/3) + i\sin(2\pi/3))]^4$

B. $[\sqrt{5}(\cos(7\pi/6) + i\sin(7\pi/6))]^8$

F. $[\sqrt[5]{7}(\cos(\pi/6) + i\sin(\pi/6))]^{10}$

C. $[\sqrt{11}(\cos(4\pi/3) + i\sin(4\pi/3))]^4$

G. $[\sqrt[4]{3}(\cos(7\pi/4) + i\sin(7\pi/4))]^{16}$

D. $(3 + 3i)^5$

H. $(\sqrt{3} - i)^6$

2. Determine the indicated roots and plot them in the complex plane. Leave the roots in rectangular form.

A. the cube roots of $-8i$

C. the sixth roots of 27

B. the fourth roots of -625

D. the fourth roots of $-8 + 8\sqrt{3}i$

3. Solve each equation.

A. $z^8 = 1$

D. $z^3 = -8$

B. $z^2 = 2 - 2i\sqrt{3}$

E. $z^2 = -3\sqrt{3}/2 - (3/2)i$

C. $z^6 = -27i$

F. $z^4 = -81$

4. Solve each equation without first converting the given complex number to trigonometric form. [Hint: Substitute $a + bi$ for z and solve for a and b.

A. $z^2 = 4i$

C. $z^2 = -9i$

B. $z^2 = 5 - 12i$

D. $z^2 = 16 + 30i$

APPENDIX

Answers to Selected Exercises

Set 1
1A. 12   B. −5   C. 13   D. −12   E. 14   F. 4   G. −8   H. 30
I. 45   J. 159   2A. yes   B. no   C. no   D. yes

Set 2
1A. Let $C = \{\text{red}, \text{white}\}$.   E. Let $E = \{12, 14, 16, 18, 20, 22, 24\}$.   2A. true
B. false   C. false   D. false   E. false   F. true   3A. $\{a, b, c, d, f, g\}$
B. $\{b, d\}$ C. $\{(b, a), (b, b), (b, c), (d, a), (d, b), (d, c), (f, a), (f, b), (f, c), (g, a), (g, b), (g, c)\}$
D. $\{d, e\}$ E. $\{b, d\}$ F. $\{b\}$ G. $\emptyset$ H. $\{b\}$ I. $\{a, b, c, d, e, f, g\}$

Set 3
1A. **Q**   B. **R**   C. **Q**   D. **Z**   E. **N**   F. **Z**   G. **N**   H. **R**
I. **Q**   J. **N**   K. **Z**   L. **R**   2A. $m + 1$ B. $n - 2$ C. $c + 4$ D. $5d - 18$

Set 4
1A. $(-7, -4]$ B. $(-1, \infty)$ C. $[100, 200]$ D. $(-\infty, 9)$
E. $[5, \infty)$ F. $(-12, -10)$ G. $(-\infty, -2]$ H. $[3, 8)$
2A. $\{\, x \in \mathbf{R} \mid 2 \leq x \leq 7 \,\}$ B. $\{\, x \in \mathbf{R} \mid x < 5 \,\}$ C. $\{\, x \in \mathbf{R} \mid -1 < x \leq 100 \,\}$
D. $\{\, x \in \mathbf{R} \mid x > -8 \,\}$ E. $\{\, x \in \mathbf{R} \mid x \geq 9 \,\}$ F. $\{\, x \in \mathbf{R} \mid -10 < x < 0 \,\}$
G. $\{\, x \in \mathbf{R} \mid 6 \leq x < 17 \,\}$ H. $\{\, x \in \mathbf{R} \mid x \leq -3 \,\}$ 3A. $(7, 11]$ B. $(-11, 3]$
C. $(2, 6)$ D. $[10, 15]$ E. $(1, 10)$ F. $\emptyset$ G. $\{6\}$ H. $(-\infty, \infty)$

Set 5

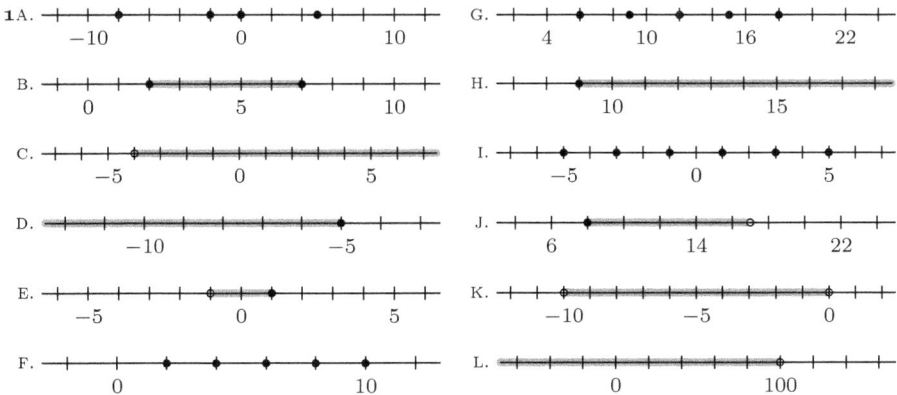

Set 6
1A. false   B. depends   C. false   D. true   E. false   F. depends
2A. yes: 4; no: −3   B. no: 0, 2, −3   C. yes: −1, 4; no: −2   D. yes: −2, 1; no: 3, 7

Set 7
1A. -8; $1/8$ B. $-2/7$; $7/2$ C. $-e$; $1/e$ D. 1; -1 E. 0; none   F. $4/3$; $-3/4$
2A. prop. of multiplicative inverses   G. commutative prop. of addition
B. commutative prop. of addition   H. associative prop. of multiplication
C. prop. of additive identity element   I. prop. of additive inverses
D. distributive property   J. prop. of multiplicative identity element
E. commutative prop. of multiplication   K. associative prop. of addition
F. prop. of additive identity element   L. commutative prop. of addition
3A. -587 B. $8{,}721$ C. $229{,}000$ D. $3{,}700$ 4. $(8 - 4) - 3 \neq 8 - (4 - 3)$
5. $6 \div 2 \neq 2 \div 6$ 6A. false   B. false   C. true   D. false

Set 8

1A. 64   B. 16   C. 25   D. -1 E. 1/125   F. 0.36
G. 1   H. -81 I. 11   J. 0   2A. $15x^3$ B. $112m^{11}$
C. $4b^5$ D. $15x^7$ 3A. -56 B. 63   C. 432   D. 13/3

Set 9

1A. $-6x$ B. $13mn^2$ C. $(203/60)z$ D. $11x^3y^5$ E. $7b + 7c$ F. $-2v + 3w$

Set 10

1A. $24x^4$ B. $4m$ C. $-2x + 37$ D. $y^3 + 4y^2 - 13y + 16$
E. $-40y^{10}$ F. $-10p + 11q$ G. $5x^2 - 20x - 67$ H. $6x - y - 25z$
I. $6x^3 + 31x^2 + 25x - 12$ J. $6m - 9n + 12mn + 6n^2 - 10$
2A. $-x - 18$; -20 B. $13y$; -104 C. 0; 0 D. $-5x^2 - 16$; -61
E. $7x - 4y - 21z - 15xy + 4xz + 6yz$; -51 F. $x^2 - y^2$; 9

Set 11

1A. additive prop. of equality   F. multiplicative prop. of equality
B. multiplicative prop. of inequality   G. transitive prop. of equality
C. substitutive prop. of equality   H. additive prop. of inequality
D. reflexive prop. of equality   I. symmetric prop. of equality
E. transitive prop. of inequality   J. multiplicative prop. of inequality

Set 12

1A. $\{11\}$ B. $\{6\}$ C. $\{-10\}$ D. $\{-4\}$ E. $\{-3/28\}$ F. $\{-3\}$ G. $\{1\}$
H. $\emptyset$ I. $\{17/60\}$ J. $\mathbf{R}$ K. $\{7\}$ L. $\{5\}$ M. $\{36\}$ N. $\{1/6\}$
O. $\{55\}$ P. $\mathbf{R}$ Q. $\{8\}$ R. $\{4\}$ S. $\{0\}$ T. $\{6\}$ U. $\emptyset$
V. $\{2\}$ 2A. 21   B. -11 C. 9   D. -3

Set 13

1A. $8 - x$ B. $m/6$ C. $2x$ D. $y + 11$ E. $p/(-3)$ F. $7 - n$
G. $x + y$ H. $14h$ I. $k + 6$ J. $12 - x$ K. $5n$ L. $x - 17$
M. $n + m$ N. $b - a$ O. $w/2$ P. $x \cdot 20$ Q. $x + 11$ R. $y - 1$
2A. $2s$ B. $b + 15$ pounds   C. $x/60$ m/s   D. $x + 2$
E. $10d$ cents   F. $t - 11$ degrees   G. $x/16$ grams   H. $m - 200$ pesos

Set 14

1A. 4   B. 18   C. \$5.25   D. 19 yards   E. 424 pages   F. 9 years old
G. 17   H. 140,101   I. 43 and 45   J. 18 feet   K. 19 nickels   L. 228 beans
M. 52 and 58 inches   N. \$84

Set 15

1A. $\{3\}$ B. $\{-2\}$ C. $\{-5/2\}$ D. $\{1\}$ E. $\{16\}$ F. $\{2\}$ G. $\{0\}$ H. $\{1/2\}$
I. $\{-2\}$ 2A. 11, 13   B. 7, 10   C. 17   D. 7   E. 17, 19   F. 26, 28, 30

Set 16

1A. $\{20/3\}$ B. $\{63/2\}$ C. $\{7/3\}$ D. $\{6/5\}$ E. $\{17/23\}$ F. $\{85\}$
G. $\{16/9\}$ H. $\{-153/44\}$ I. $\{118/15\}$ J. $\{-111/14\}$ K. $\emptyset$ L. $\mathbf{R}$
2A. 9/2   B. 29/7   C. $6°$ D. $123°$

Set 17

1A. $\{22\}$ B. $\{-9\}$ C. $\{4\}$ D. $\{4/11\}$ E. $\{10/21\}$ F. $\{-1\}$
G. $\{-38\}$ H. $\{99\}$ I. $\{-1/9\}$ J. $\{-78/7\}$ 2A. 212   B. 0
C. 59   D. 30   E. -40 3A. 5/9   B. 139/333   C. 686/99
D. 1037/300   E. 1/45   F. 875/99   G. 7186/9999   H. 4787/2475

Set 18

1A. $(14, \infty)$ B. $(-\infty, -2]$ C. $[3, \infty)$ D. $(-\infty, 44)$ E. $(-\infty, -20)$ F. $(19, \infty)$
G. $(-2, 5]$ H. $[-4, \infty)$ I. $(-\infty, 4)$ J. $(-\infty, -3)$ K. $(-\infty, -35]$ L. $[-1, \infty)$
M. $(-\infty, -6]$ N. $(-5, 2]$ 2A. $(4, \infty)$ B. 98   C. \$19   D. 22 games

Set 19

1A. $[-5, 3)$ B. $(-\infty, 1) \cup (4, \infty)$ C. $\mathbf{R}$ D. $\emptyset$ E. $(-\infty, -3]$
F. $(-\infty, -2]$ G. $(8, \infty)$ H. $[9, \infty)$ 2A. $(-\infty, 2)$ B. $\emptyset$ C. $(-5, 12]$
D. $(-\infty, 3) \cup [8, \infty)$ E. $(-1, \infty)$ F. $\mathbf{R}$ G. $(-\infty, -3]$ H. $(3, \infty)$
3A. $[59, 99)$ B. $[0, 57) \cup [97, 100]$

Set 20

1A. 10   B. 0   C. 7   D. $-\pi$ 2A. $\{-8, 8\}$ B. $\{0\}$
C. $\{-3.3\}$ D. $\{-16, 16\}$ E. $\{-30, 30\}$ F. $\{-1/2, 2\}$ G. $\{-21, 9\}$ H. $\{-1, 7\}$
I. $\emptyset$ J. $\{-3, 3\}$ K. $\{-6, 6\}$ L. $\{-2, 2\}$ M. $\{-16, 16\}$ N. $\{-13, 49/3\}$
O. $\{-17/2, 9\}$ P. $\{-13/5, 3\}$ 3A. $\emptyset$ B. $[-5, 5]$ C. $(-\infty, -4) \cup (4, \infty)$
D. $(-\infty, -6] \cup [6, \infty)$ E. $[-21, 21]$ F. $(-\infty, -1) \cup (7/9, \infty)$ G. $(-\infty, -13/2) \cup (6, \infty)$
H. $(-7/3, 3)$ I. $\mathbf{R}$ J. $(\infty, 0) \cup (0, \infty)$ K. $[-7, 7]$ L. $(-21, 21)$
M. $(-\infty, 1] \cup [8, \infty)$ N. $(-36, 36)$ O. $[-5, 27]$ P. $(-\infty, -3/2) \cup (5, \infty)$

Set 21

1A. d: 4; m: 7   B. d: 13; m: 9/2   C. d: 6; m: -4 D. d: 10; m: 5
E. d: 1; m: $-11/2$ F. d: 8; m: 11   G. d: 13; m: $-13/2$ H. d: 6; m: 0

Set 22

1A. A: 40; P: 26   B. A: 49π cm^2; C: 14π cm   C. V: $256\pi/3$ ft^3; S: 64π ft^2
D. V: 24π in^3; S: 32π in^2 2A. 99.2 in   B. 405 cm^3 C. 150π in^3 D. 240 yd
E. 1,500 ft^2 F. 9π cm^2 G. 18π km   H. 384 in^3 I. $361/4$ in^2 J. 114 cm^2
3A. 40   B. 24   C. 100   D. 18   E. 63   F. 25
4A. $b = A/h$ B. $h = 3V/(\pi r^2)$ C. $P = A/(1 + rt)$ D. $w = (P - 2l)/2$
E. $l = (S - 2wh)/(2w + 2h)$ F. $C = (5/9)(F - 32)$ 5. $(x - 3)/(5 + 2x)$

Set 23

1A. $\{21\}$ B. $\{-3/11\}$ C. $\{-69/29\}$ D. $\{1/16\}$ E. $\{1/42\}$ F. $\{9\}$
G. $\{11\}$ H. $\{-4/5\}$ I. $\{51/32\}$ J. $\{2\}$ 2A. 896 lbs   B. 2,160 mi
C. 63 in   D. $77/2$ ft   E. 16   F. 280   G. 292.5 m   H. 45 m

Set 24

1A. $d = kw$ B. $P = k/V$ C. $V = kr^3$ D. $l = kw$ E. $A = kbh$ F. $c = k/d$
G. $x = ky/z^2$ 2A. 4   B. 240   C. $1/\pi$ D. 2π
3A. 6   B. 11   4A. 8   B. 2   5. 72   6. 8   7. 19   8. 5

Set 25

1A. 0.78   B. 3.50   C. 0.08   D. 0.001   2A. 35%   B. 0.087%
C. 5,263%   D. 4%   3A. 60%   B. 175%   C. 2%   D. 66.$\overline{66}$%
4A. 9/5   B. 59/100   C. 1/4   D. 4   5A. 11   B. 63
C. 60   D. 450   E. 35%   F. 141%   G. 126   H. 0.37   I. 375   J. 54
K. 76%   L. 0.3%   6A. 5%   B. \$4,725   C. 2,812   D. 20 cm   E. 14   F. 14%

Set 26

1A. \$2.40   B. \$3.50   C. 16%   D. \$2,318   E. \$81,900   F. 35%
2A. \$13.33   B. \$4.80   C. 175%   D. \$5.34   E. \$2.00   F. 45%
3A. 32%   B. 52.88%   C. increase of 3.5%

Set 27

1A. \$800   B. 27%   C. 87 years old   D. \$12,800   E. \$2,000
F. \$11,500 at 2.2%; \$35,000 at 5.7%   G. 0.2%   H. 40 years   I. \$60,000
J. \$2,100 at 12%; \$7,900 at −7%   K. 50 years

Set 28

1A. 20 L   B. 12%   C. 18.75 gal   D. 45 mL   E. 12 qt of 35%; 18 qt of 60%
F. 15 g   G. 20%   H. 6,000 gal   I. 48%   J. 32 oz of old; 16 oz of new

Set 29

1A. d: 2; l.c.: 9   B. d: 0; l.c.: −8   C. d: 3; l.c.: −5   D. d: 2; l.c.: 10
E. d: 1; l.c.: 1   F. d: 8; l.c.: 3   G. d: 5; l.c.: −7   H. d: 2; l.c.: 1
I. d: 0; l.c.: 19   J. d: 3; l.c.: −4   K. d: 5; l.c.: 4   L. d: 1; l.c.: −1
2A. $9x^2 - 4x - 11$ B. $6x^2 + 12x - 6$ C. $5x - 7m - 2$ D. 0
E. $8x - 6$ F. $5x + 8y - 2$ G. $8b^2 + 3b + 4$ H. $x^2 + 2x + 9$
I. $-13x^3 + 6x^2 - x + 3$ J. $2x^3 - 15x^2 + 2x - 1$ K. $-5k^4 + 4k^3 + 2k^2 - 2k + 17$
L. $-9n^3 + 6n^2 - 6n - 7$ M. $w^3 - w^2 + 2w - 1$ N. $2a^2 - 3b^2 - 2a + 5b + 8$

Set 30

1A. $24x^9$ B. $-16x^7y^6$ C. $(-12/5)u^6v^8$ D. $49x^8$
E. $a^{90}b^{50}c^{20}$ F. $24x^2 + 21x$ G. $288x^{13}y^{24}$ H. $-30t^9 + 40t^8 + 90t^7$
I. $27w^6$ J. $28p^4q^{14}$ K. $2x^{10}y^7$ L. $-27m^6n^3$
M. $243w^5x^{10}y^{15}z^{20}$ N. $4w^3 - 8w^4 - 4w^5$ O. $-x - x^2 - x^3$ P. $m^3n - m^2n^2 + mn^5$
Q. $x^5y^9 + x^3y^{10} - xy^{11}$ R. $200x^{17}y^9z^7$ S. $49x^4 + 14x^3 - 12x^2$
T. $-83x^2 - 18x$ U. $-5x^3 - 2x^2 + 57x + 2$ V. $-38x^3$

Set 31

1A. $x^2 + 10x + 16$ B. $4n^2 + 31n - 45$ C. $18b^2 - 9b - 14$ D. $cd + 3c - 2d - 6$
E. $x^3 - 1$ F. $2x^4 + 3x^3 - 4x^2 + 6x - 16$ G. $m^3 + 6m^2 + 12m + 8$
H. $x^2 + 6x - 7$ I. $18m^2 - 51m + 15$ J. $3x^2 - 26x - 40$ K. $vw + 10v + 5w + 50$
L. $12x^4 - 4x^3 - 63x^2 - 53x - 12$ M. $y^3 + 27$ N. $8k^3 - 36k^2 + 54k - 27$
2A. $x^2 - 10 + 25$ B. $9m^2 + 42m + 49$ C. $36 - 24m + 4m^2$ D. $81x^2 + 72xy + 16y^2$
E. $x^2 - 9$ F. $49 - 25k^2$ G. $1 - x^4$ H. $9m^2 - n^2$
I. $y^2 + 8y + 16$ J. $25x^2 - 90x + 81$ K. $100 - 20x + x^2$ L. $4a^2 - 28ab + 49b^2$
M. $m^2 - 64$ N. $16x^2 - 81$ O. $4x^4 - 16$ P. $64x^2 - 25y^2$

Set 32

1A. $24x^2 + 14x + 2$ E. $8x^2 - 2x - 45$
B. $14a^2 - 4b^2 - 30c^2 + 26ab + 64ac - 22bc$ F. $5x^2 - 12y^2 + 16z^2 - 11xy - 42xz - 26yz$
C. $x^3 - 2x^2 - 4x + 8$ G. $c^3 - 2c^2 - 21c - 18$
D. $6x^3 + 19x^2 - 51x + 20$ H. $15x^3 - 97x^2 - 212x + 224$
2A. 64   B. 29

Set 33

1A. $x^2 - 11x + 30$ B. $2n^2 - 3n - 27$ C. $-3p^3 - 7p^2 - 12p - 28$
D. $20r^2 + 13r - 72$ E. $28c^5 - 5c^4 - 3c^3$ F. $x^2 - 9x - 10$ G. $4m^2 + 33m + 8$
H. $55x^2 + 54x + 8$ I. $-36d^2 + 12d - 1$ J. $2q^{11} - 5q^8 - 12q^3 + 30$

Set 34

1A. x^5 B. $1/w^{13}$ C. $1/(2v)$ D. $9n^8$ E. $-4b^4/a^8$ F. $-u^2/(3w^4)$
G. $-x^2b^2/2$ H. $-3x - 7x^7y$ I. $-b^3/(6a) + b^4/(2a^2)$ J. $y^2/25$ K. $16x^{20}/y^{12}$
L. $r^{18}t^{90}/s^{45}$ M. $1/y^7$ N. 1   O. $-9t^7$ P. $-7/(5x^{15})$ Q. $-y/(4x)$
R. $5/(v^3w^3)$ S. $-st/(2r^7)$ T. $-4m^5n^5/3 - 2m^4n^3$ U. $2x^2y^3 - 13/(5y^2)$
V. $b^3/(27a^6)$ W. $a^{16}/(b^{56}c^{24})$ X. $81w^{14}/(u^8v^4)$

Set 35

1A. q: $x - 9$; r: -3 B. q: $2m - 3$; r: -11 C. q: $2x^2 - 1$; r: -6 D. q: $(1/2)x - 3$; r: 13
E. q: $-v^2 - 7$; r: 1   F. q: $y^3 + 3y^2 + 9y + 27$; r: 0   G. q: $x - 1$; r: 13
H. q: $-5b - 12$; r: 0   I. q: $-2x - 3$; r: 7   J. q: $x^4 - x^3 + x^2 - x + 1$; r: 1
K. q: $3c^3 - 6c^2 + 2c$; r: 0   L. q: $(3/2)z^2 + 2z + 1$; r: -2 M. q: $9x - 1$; r: $36x - 4$
N. q: $3x - 4$; r: 0   O. q: $x^2 + 5x + 3$; r: 0   P. q: $5x^2 + 8x - 9$; r: $-46x$
2A. -15 B. 15   C. -5 D. -8 E. 7   F. -14 3. $b = 5$; $c = 49$

Set 36

1A. q: x; r: -7 B. q: $3b - 2$; r: 10   C. q: $x^2 - 4x + 16$; r: 0
D. q: $6n + 8$; r: 7   E. q: $k^2 - 2k - 2$; r: -15 F. q: $-y + 1$; r: -3
G. q: $4x^2 + x + 7$; r: 44   H. q: $y + 7$; r: -26 I. q: $2a + 3$; r: 4
J. q: $-m^4 - 2m^3 - 4m^2 - 8m - 16$; r: 0   K. q: $-5x^2 - 2x - 14$; r: -9
L. q: $-12x - 3$; r: -3 M. q: $v^2 + 2v + 1$; r: 3   N. q: $3x^2 - 6x + 3$; r: -11
O. q: $x^3 + 2x^2 + 4x + 1$; r: 20   P. q: $5b^4 - 3b^3 + 6b^2 - 16b + 31$; r: -51

Set 37

1A. $1/64$ B. 1   C. 81   D. $32/243$ E. 9,261   F. 2,916
G. $1/8$ H. 49   I. 1   J. $125/8$ K. $1/4,096$ L. 160,000
2A. $1/x^{14}$ B. $21m^5$ C. $-w^7/6$ D. $-3/b^{19}$ E. $-1/128$ F. $36/(25v^8)$
G. 1   H. $b^9/(4a^9)$ I. $1/y^4$ J. $-36/n^6$ K. $16/(9x^3)$ L. $m^5/11$
M. $-t^{24}/64$ N. 256   O. $2/u^{15}$ P. $3/(r^4s^5)$

Set 38

1A. $18 = 2 \cdot 3^2$; $32 = 2^5$; GCF $= 2$ B. $35 = 5 \cdot 7$; $72 = 2^3 \cdot 3^2$; GCF $= 1$
C. $24 = 2^3 \cdot 3$; $36 = 2^2 \cdot 3^2$; $90 = 2 \cdot 3^2 \cdot 5$; GCF $= 6$
D. $40 = 2^3 \cdot 5$; $75 = 3 \cdot 5^2$; GCF $= 5$ E. $56 = 2^3 \cdot 7$; $71 = 71$; GCF $= 1$
F. $28 = 2^2 \cdot 7$; $84 = 2^2 \cdot 3 \cdot 7$; $140 = 2^2 \cdot 5 \cdot 7$; GCF $= 28$
2A. $6y$ B. $9m^3n^2$ C. $2x^2y^8z$ D. $2y^5$ E. $11ab^2$ F. $13q^3r^7st^9$
3A. $4x^2(2x^2 + 3)$ B. $xy(9x - 17)$ C. $2t^3(2t^2 - 6t + 3)$ D. $(6x - 5)(x - 2)$
E. $2y(6y + 5)(7 - 4z)$ F. $(6x - 5)(x - 2)$ G. $10y(3 - 4y^7)$ H. $3m^5n^3(19m^3 + 18n^7)$
I. $7u(2 - 5u - 7u^3)$ J. $(m + 2n)(3c + 4)$ K. $(x + 8)(11 - 7x)$ L. $5x^6(4 - 11x^2)(x - 3)$

Set 39

1A. $(x + y)(c + 5)$ B. $(r^2 - 2s)(7 - k)$ C. $(2m - 1)(5m + 4)$
D. $(3x + 1)(5 - 6x^2)$ E. $(5x - 1)(20x^2 + 3)$ F. $(x + y)(a - 1)$
G. $x^2(3x^2 + 4)(3x^2 - 4)$ H. $(5x - 2y + 3z)(2 - b)$ I. $(m - n)(b + 3)$
J. $-(u + v)(1 + t)$ K. $(3y + 2)(12y + 7)$ L. $(4n - 1)(3n^2 - 1)$
M. $-4(1 + 6x)(1 + 2x^2)$ N. $(2c^2 - 5)(2c^2 + 3)$ O. $(8 - t)(u - 2v)$
P. $(x^2 + x - 8)(7x^2 + 5)$ 2. 130

Set 40

1A. $(x - 3)(x + 3)$ B. $(n - 9)(n + 9)$ C. $(u - v)(u + v)$ D. $2(w - 6)(w + 6)$
E. $(2x - 5y)(2x + 5y)$ F. $2s^2t^3(4t - 7)(4t + 7)$ G. $x^2(x - 1)(x + 1)$
H. $(x + 2)(2x - 5)(2x + 5)$ I. $(m - 8)(m + 8)$ J. $(x - 11)(x + 11)$
K. $(x - y)(x + y)$ L. $(7s - 10t)(7s + 10t)$ M. $3u(3 - 2v)(3 + 2v)$
N. $15x(x - 1)(x + 1)$ O. $3(1 - 4y^3)(1 + 4y^3)$ P. $3(x - 2)(x - 3)(x + 3)$

Set 41

1A. $(x - 3)(x^2 + 3x + 9)$ B. $5(2 + b)(4 - 2b + b^2)$ C. $(5 + 4m)(25 - 20m + 16m^2)$
D. $(y^2 + 7z)(y^4 - 7y^2z + 49z^2)$ E. $(2r - 4st)(4r^2 + 8rst + 16s^2t^2)$
F. $(1 - u)(1 + u + u^2)(1 + u^3 + u^6)$ G. $(x + 2)(x^2 - 2x + 4)$
H. $(1 - 6n)(1 + 6n + 36n^2)$ I. $4n(1 + 3m)(1 - 3m + 9m^2)$
J. $(ab + c^4)(a^2b^2 - abc^4 + c^8)$ K. $(x - 1)(x + 1)(x^2 + 1)(x^8 + x^4 + 1)$
L. $(x - 2y)(x + 2y)(x^2 + 2xy + 4y^2)(x^2 - 2xy + 4y^2)$ 2. 80

Set 42

1A. $(x+2)(x+3)$ B. $(y+10)(y-2)$ C. $(m-5)(m-8)$ D. $(n-10)(n+6)$
E. $(c+6)(c+3)$ F. $(x-4)(x+2)$ G. $(v-8)(v-3)$ H. $(w+6)(w-2)$
I. $(x-8)(x+7)$ J. $(b-10)(b-7)$ K. $(k+5)(k+2)$ L. $(x+5)(x-4)$
M. $(a-9)(a+7)$ N. $(x+9)(x+6)$ O. $(x-2)(x-1)$ P. $(a-5)(a+1)$
Q. $(b+7)(b-3)$ R. $(y-9)(y-5)$ S. $(w+7)(w+4)$ T. $(v+10)(v-8)$
U. $(t-3)(t+1)$ V. $(s+10)(s+4)$ W. $(r-10)(r-1)$ X. $(x-9)(x+4)$
Y. $(c+8)(c-4)$ Z. $(t-4)(t+3)$ 2A. $(y^2+6)(y^2+4)$ B. $3(m+9)(m-1)$
C. $m(n-5)(n-3)$ D. $5y^3(x+3)(x-2)$ E. $(x-10y)(x-8y)$ F. $(y-7)(y+7)$
G. $4(m-5)(m+5)$ H. $(a^3+7)(a^3-5)$ I. $2(v-7)(v-2)$ J. $c(b+10)(b+5)$
K. $3s(t-10)(t+3)$ L. $(m+8n)(m-2n)$ M. $(k-1)(k+1)$ N. $7(c-2)(c+2)$

Set 43

1A. $(2x-3)(x+4)$ B. $(3y-1)(y-5)$ C. $(2a-9)(2a+3)$ D. $(6n+5)(n+2)$
E. $2(3w-7)(w-6)$ F. $5k(2k+7)(2k-1)$ G. $(2x+7)(x+9)$ H. $(3z+4)(z-6)$
I. $(4m-5)(m-6)$ J. $(3b-10)(2b+11)$ K. $3(5z+2)(z-8)$ L. $4s(9r+40)(r-1)$

Set 44

1A. $\{-7,0\}$ B. $\{0,10\}$ C. $\{-4,4\}$ D. $\{-3,-4\}$ E. $\{-2,8/3\}$ F. $\{-9,-8\}$
G. $\{-1,7\}$ H. $\{5,6\}$ I. $\{6,7\}$ J. $\{-8,-1\}$ K. $\{-1,0,1\}$ L. $\{3,9/2\}$
M. $\{-2,0\}$ N. $\{0,3\}$ O. $\{-8,0\}$ P. $\{-7,7\}$ Q. $\{3,9\}$ R. $\{-8,6\}$
S. $\{1/2,-5/2\}$ T. $\{14/5,9/2\}$ U. $\{-4,-1\}$ V. $\{-2,4\}$
W. $\{-9,2\}$ X. $\{-3,0,3\}$ Y. $\{-2,0,9\}$ Z. $\{-5/3,1/2\}$
2A. w: 4 ft; l: 6 ft   B. b: 7 in; h: 10 in   C. 9 years old   D. 13 marbles

Set 45

1A. $7y/5$ B. $3m/n^2$ C. $-2/(3ab^4c^2d)$ D. $1/(3x)$
E. $(x-4)/(x+6)$ F. $(x+8)/(x-2)$ G. $(3x+5)/(x-1)$ H. $-(2x+5)/(3x)$
I. $(5x-8)/(x^2+4x+16)$ J. $-2/(5b)$ K. $-2/(a^2b^2)$
L. $3x^2y^7z$ M. $x/3$ N. $(x-3)/(x+4)$ O. $(3x-2)/(5x+6)$
P. $(3x-10)/(7x+1)$ Q. $-(3x+4)/(2x+9)$ R. $(4x^2-6x+9)/(x-11)$

Set 46

1A. 10 B. $-y/5$ C. $8x$ D. $(2x+1)/(x-4)^2$
E. 1 F. $15/4$ G. $8c^4/(a^4b)$ H. $7m(m+5)/(2n+4)$
I. $(x-2)(x+1)/[2(3x-5)(x+2)]$ J. $(c+7)^2/[(c+2)(2c-9)]$

Set 47

1A. $(7x+1)/6$ B. $(17n-12)/15$ C. $6/p$
D. $(11x-37)/[(x+1)(x-5)]$ E. $(17x+21)/(28x^2)$ F. $(10x^2-3x+3)/[4x(x-1)]$
G. $(15m-3m^2-4n-6n^2)/(30mn)$ H. $(13x-10)/[x(x-2)(x+5)]$
I. $(-12x^2+96x-46)/[(x+1)(x-2)(x-10)]$
J. $(5x^2-28x+60)/[(x-4)(x-8)(2x+5)]$
K. $(-10n^2+13n+24)/[(n+2)(3n+13)(n^2-2n+4)]$
L. $(2-17m)/5$ M. $(9x-5)/4$ N. $3/2$
O. $(-7x+12)/[x(2x-3)]$ P. $(6x^2-9x-12)/(8x^2)$ Q. $(-7x+29)/(18x-12)$
R. $(2a-24c+ab+8bc)/(8abc)$ S. $(-14x+1)/[(x-1)(x-4)(x+7)]$
T. $(3x^2-47x+28)/[(x+7)(x+3)(x-5)]$ U. $(2x^2+35x+1)/[(x+9)(2x-1)(3x-7)]$
V. $(3m^2+2m+90)/[(m-3)(2m-5)(m^2+3m+9)]$
W. $6/[x(x-2)(x+2)]$ X. $(20x^2-5x-30)/[(x-5)(x+1)(x+6)]$

Set 48

1A. $14/5$ B. $135/658$ C. $(x^5+y^2)/(x^5y-x^2y^2)$
D. $(x+5)(x+7)/3$ E. $5(x+2)/4$ F. $-3/[(x+2)(x+h+2)]$
G. $2(x+1)(x+7)/[(x-2)(x+4)]$ H. $(4x+19)/(2-x)$

I. $4x$

L. $(9x - 45)/(2x - 16)$

O. $(-2x - 4)/(x + 6)$

J. $387/70$

M. $7x/12$

P. 5

K. $(x^6 y^4 + x^7 y)/(y^4 - x^7)$

N. $(-16x - 8h)/[x^2(x + h)^2]$

Set 49

1A. $\{-15/2\}$ B. $\{-4/9\}$ C. $\{-135/16\}$ D. $\{-168/43\}$ E. $\{9/2\}$ F. $\{2\}$

G. $\{-1, 0\}$ H. $\emptyset$ I. $\{80/29\}$ J. $\{2/35\}$ K. $\{-17/2\}$ L. $\{3\}$

M. $\{1, 10\}$ N. $\emptyset$ O. $\{-5, 2\}$ P. $\emptyset$ 2A. -12 OR -7 B. $-8/7$ OR 3

Set 50

1A. 21 hr   B. 35 km   C. 33 km/hr, 41 km/hr   D. 34.5 s

E. 0.5 mi   F. 4.5 mi   G. 24 km/hr   H. 6 mi/hr

Set 51

1A. $3/2$ hr $= 1$ hr 30 min   B. $400/21$ min ≈ 19 min 3 s   C. $253/12$ min $= 21$ min 5 s

D. 32   E. 40 min   F. 41   G. 29   H. $56/3$ min $= 18$ min 40 s

I. $9/2$ hr $= 4$ hr 30 min   J. $11/2$ hr $= 5$ hr 30 min   K. 420   L. 1,232

Set 52

1A.
product	$+$	$-$	$+$
$x + 1$	$-$	$-$	$+$
$x + 5$	$-$	$+$	$+$
	-5	-1	

B.
product	$+$	$-$	$+$
$2x + 5$	$-$	$+$	$+$
$5x - 4$	$-$	$-$	$+$
	$-\frac{5}{2}$	$\frac{4}{5}$	

C.
product	$+$	$-$	$+$
$4 - x$	$+$	$+$	$-$
$5 - 3x$	$+$	$-$	$-$
	$\frac{5}{3}$	4	

D.
expression	$+$	$-$	$+$
$x - 9$	$-$	$-$	$+$
$x + 3$	$-$	$+$	$+$
	-3	9	

E.
product	$+$	$-$	$+$
3	$+$	$+$	$+$
$x - 5$	$-$	$+$	$+$
$x - 7$	$-$	$-$	$+$
	5	7	

F.
product	$+$	$-$	$+$	$+$
x	$-$	$-$	$+$	$+$
$(x - 4)^2$	$+$	$+$	$+$	$+$
$(x + 8)^3$	$-$	$+$	$+$	$+$
	-8	0	4	

G.
quotient	$+$	$-$	$+$
$3x - 5$	$-$	$+$	$+$
$x - 12$	$-$	$-$	$+$
	$\frac{5}{3}$	12	

H.
quotient	$-$	$-$	$+$	$-$	$+$
$-8x$	$+$	$+$	$-$	$-$	$-$
$4 - x$	$+$	$+$	$+$	$-$	$-$
$(2x - 9)^3$	$-$	$-$	$-$	$-$	$+$
$(x + 6)^4$	$+$	$+$	$+$	$+$	$+$
	-6	0	4	$\frac{9}{2}$	

I.
product	$+$	$-$	$+$
$x - 7$	$-$	$-$	$+$
$x + 4$	$-$	$+$	$+$
	-4	7	

J.
product	$+$	$-$	$+$
$7x - 1$	$-$	$+$	$+$
$8x - 3$	$-$	$-$	$+$
	$\frac{1}{7}$	$\frac{3}{8}$	

K.
product	$+$	$-$	$+$
$11 - 2x$	$+$	$-$	$-$
$10 - x$	$+$	$+$	$-$
	$\frac{11}{2}$	10	

L.
expression	$+$	$-$	$+$
$x - 8$	$-$	$-$	$+$
$x - 2$	$-$	$+$	$+$
	2	8	

M.

product	−	+	−
−4	−	−	−
2x + 1	−	−	+
x + 8	−	+	+

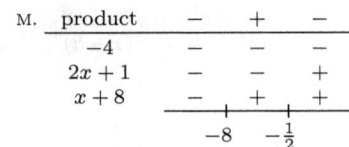

$-8 \qquad -\frac{1}{2}$

N.

product	+	−	−	−
−9x²	−	−	−	−
(x + 1)⁵	−	+	+	+
(x − 99)⁴⁰	+	+	+	+

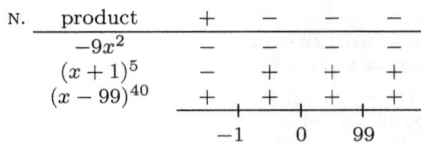

$-1 \qquad 0 \qquad 99$

O.

quotient	+	−	+
x + 2	−	+	+
8x + 1	−	−	+

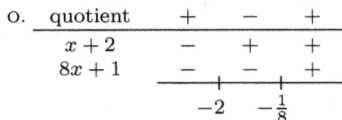

$-2 \qquad -\frac{1}{8}$

P.

quotient	−	−	+	−	+
−x³	+	+	+	−	−
(x + 3)²	+	+	+	+	+
(3x + 7)⁵	−	−	+	+	+
1 − 6x	+	+	+	+	−

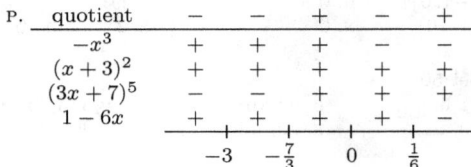

$-3 \qquad -\frac{7}{3} \qquad 0 \qquad \frac{1}{6}$

2A. expression

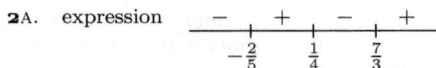

$-\frac{2}{5} \qquad \frac{1}{4} \qquad \frac{7}{3}$

B. expr.

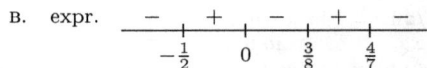

$-\frac{1}{2} \qquad 0 \qquad \frac{3}{8} \qquad \frac{4}{7}$

C. expression

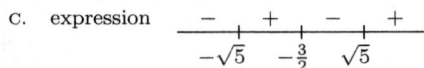

$-\sqrt{5} \qquad -\frac{3}{2} \qquad \sqrt{5}$

D. expression

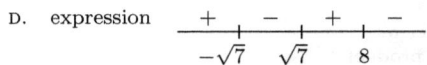

$-\sqrt{7} \qquad \sqrt{7} \qquad 8$

Set 53

1A. $(-\infty, -1] \cup [7, \infty)$ B. $(5/2, 5)$ C. $(-10, -3) \cup (0, \infty)$
D. $(-\infty, -3] \cup [-2/5, 3]$ E. $(-\infty, -10) \cup (2, \infty)$ F. $(-\infty, -3/2] \cup (8, \infty)$
G. $(-\infty, 0) \cup [1, 2]$ H. $(-3, -1/5) \cup (2, 8/3)$ I. $(-\infty, -4) \cup (-4, 0] \cup (5, \infty)$
J. $(-\infty, -2) \cup (-17/12, \infty)$ K. $(-\infty, -8) \cup (-2, 2) \cup (4, \infty)$
L. $(-\infty, -5) \cup (-2, -1/3] \cup [3, \infty)$ M. $[-13, -5) \cup (-5, -7/3]$ N. $[-4, -3]$
O. $(-\infty, -3) \cup (4/3, \infty)$ P. $[-6, 0] \cup [12, \infty)$ Q. $(-\infty, -2) \cup (1/7, 2)$
R. $[5, 6]$ S. $(-15, -3)$ T. $(-7, 4) \cup (5, \infty)$
U. $(-\infty, -11) \cup [-7/2, -1) \cup [9, \infty)$ V. $(-8, 3) \cup (3, 6)$ W. $(-5, -9/5]$
X. $(-\infty, 3/2] \cup (3, 4) \cup [5, \infty)$ Y. $(-8, -7) \cup (4, 9/2)$ Z. $(-\infty, 1] \cup [9, \infty)$

Set 54

1A. 7 B. −4 C. 4 D. −1 E. 2/3 F. 2
G. 1 H. 0.2 I. not real J. 3 K. −8 L. 3
M. −2 N. 11/5 O. 3 P. −2 Q. 0.5 R. not real
2A. $7\sqrt{3}$ B. $-5\sqrt[3]{6}$ C. $3\sqrt{5}$ D. $-6\sqrt{17} - 7\sqrt[4]{17}$ E. $7\sqrt{5} - 5\sqrt{2}$
F. $8\sqrt{2}$ G. $-7\sqrt[4]{7}$ H. $14\sqrt{11}$ I. $-8\sqrt{13} - 11\sqrt[3]{13}$ J. $16\sqrt{3} - \sqrt{10}$

Set 55

1A. 7 B. 39 C. −5 D. 10 E. 13 F. 2 G. 3 H. 211
I. 6 J. 23 K. 71 L. −5 2A. $4\sqrt{3}$ B. $3\sqrt{6}$ C. $2\sqrt[3]{10}$ D. $-4\sqrt[3]{3}$
E. $\sqrt{5}/9$ F. $\sqrt{3}/7$ G. $-\sqrt[3]{2}/3$ H. $\sqrt[5]{13}/2$ I. $5\sqrt{7}$ J. $2\sqrt{5}$ K. $-3\sqrt[3]{4}$ L. $5\sqrt[3]{2}$
M. $\sqrt{11}/6$ N. $\sqrt{7}/8$ O. $\sqrt[4]{5}/3$ P. $-\sqrt[3]{17}/2$ 3. $5 - 2x$ 4A. positive B. negative

Set 56

1A. $2\sqrt{3}$ B. $3\sqrt{6}$ C. $36\sqrt{5}$ D. $2\sqrt[3]{7}$ E. $x^2\sqrt{x}$ F. $n^5\sqrt[3]{n^2}$
G. $k^3\sqrt[8]{k}$ H. $5m^7\sqrt{3m}$ I. $-3xy^2z^3\sqrt[3]{2xz^2}$ J. $7\sqrt{2}$ K. $10\sqrt{5}$
L. $35\sqrt{3}$ M. $-3\sqrt[3]{3}$ N. $m^5\sqrt{m}$ O. $c^3\sqrt[6]{c^5}$ P. $t\sqrt[5]{t}$ Q. $6x\sqrt{10x}$
R. $2x^3yz^5\sqrt[4]{5y^3z}$ 2A. $3\sqrt{10}/10$ B. $2\sqrt[3]{3}/3$ C. $4\sqrt{5x}/x^2$ D. $\sqrt{35}/10$
E. $5\sqrt[6]{b^5c^5}/(bc^3)$ F. $m\sqrt[3]{mn}/n^4$ G. $5\sqrt{3}/3$ H. $11\sqrt[3]{4}/2$
I. $2n^3\sqrt{7mn}/m^2$ J. $3\sqrt[5]{k^3m}/(km^3)$ K. $\sqrt{10}/6$ L. $m^2/\sqrt[7]{m^3n^6}/n$
3A. $-\sqrt{2}$ B. $2\sqrt{5}$ C. 12/5 D. 35/9 E. $15\sqrt{3}$ F. 1/6 G. 2 H. $-29\sqrt{10}$

Set 57

1A. 4

B. $3\sqrt{10}$

C. $42\sqrt{15}$

D. $-45\sqrt{2} - 315\sqrt{3}$

E. $7\sqrt{2} - 7\sqrt{3} + 2\sqrt{7} - \sqrt{42}$

F. $102\sqrt{2} + 510\sqrt{5}$

G. $2\sqrt[3]{4}$

H. $a^4 b^5$

I. $21 + 2x - 17\sqrt{x}$

J. $\sqrt{5}\sqrt[3]{10} - 5\sqrt{2}$

K. 6

L. $\sqrt{22}$

M. $-56\sqrt{2}$

N. $-12 + 15\sqrt{2}$

O. $2\sqrt{6} - 2\sqrt{15} - \sqrt{22} + \sqrt{55}$

P. 523

Q. 3

R. $k^2 n^3 \sqrt{kn}$

S. $8 - 18m^7$

T. $\sqrt{8}\sqrt[3]{4} + 2\sqrt[3]{6}$

Set 58

1A. $\sqrt{15}/5$

B. $\sqrt{102}/6$

C. $\sqrt{5}/10$

D. $1/3$

E. 20

F. $10\sqrt{3m}/(3m)$

G. $9x^3\sqrt{x}$

H. $2\sqrt[3]{4c^2}/c^3$

I. $7\sqrt{10}/5$

J. $(5 - \sqrt{3})/11$

K. $(-24 - 8\sqrt{2})/7$

L. $(7 + 3\sqrt{7})/4$

M. $(24 + x - 11\sqrt{x})/(64 - x)$

N. $\sqrt{14}/2$

O. $\sqrt{10}/5$

P. $\sqrt{7}/35$

Q. $1/2$

R. 33

S. $6v\sqrt{5w}/(5w)$

T. $2\sqrt{x}/x^3$

U. $m\sqrt[5]{m^4 n^3}/n^4$

V. $17\sqrt{7}/7$

W. -20

X. $7 - 3\sqrt{5}$

Y. $(103 - 20\sqrt{3})/97$

Z. $(18n + 20 + 53\sqrt{n})/(81n - 16)$

2A. $-35/(56 - 21\sqrt{11})$

B. $1/(3\sqrt{15} + 3\sqrt{11})$

C. $60/(11\sqrt{3})$

D. $1/(117 + 52\sqrt{5})$

E. $2/(9\sqrt{2} - 6\sqrt{3})$

F. 0

Set 59

1A. $\{9\}$

B. $\emptyset$

C. $\{-5, -3\}$

D. $\{2/3, 6\}$

E. $\{8\}$

F. $\{10/9\}$

G. $\{7\}$

H. $\{6\}$

I. $\{5\}$

J. $\{1\}$

K. $\emptyset$

L. $\{2, 7\}$

M. $\{-10, 8\}$

N. $\emptyset$

O. $\{3/2\}$

P. $\{3\}$

Q. $\{9\}$

R. $\{1\}$

2A. $\{4, 12\}$

B. $\{2, 6\}$

C. $\{10\}$

D. $\{-13/25\}$

3A. $\{7\}$

B. $\{-10, 11\}$

C. $\{-5, -3, 3\}$

D. $\{1, 8\}$

E. $\{-16\}$

F. $\{4, 9\}$

G. $\{-4, 3, 4\}$

H. $\{4/5, 5\}$

Set 60

1A. $\{-3, 3\}$

B. $\{5\}$

C. $\{-4, 4\}$

D. $\{-2\}$

E. $\{-2\sqrt{6}, 2\sqrt{6}\}$

F. $\{-9/2, 9/2\}$

G. $\{1, 9\}$

H. $\{-2, 4/3\}$

I. $\{1, 5\}$

J. $\{-14, 10\}$

K. $\{-6, 6\}$

L. $\{-3\}$

M. $\{3\}$

N. $\{-1, 1\}$

O. $\{-7\sqrt{2}, 7\sqrt{2}\}$

P. $\{-11/5, 11/5\}$

Q. $\{-13, -9\}$

R. $\{-3/4, 17/4\}$

S. $\{-5\}$

T. $\{-3, 15\}$

2A. $\sqrt[3]{7}$

B. $1/4$

3A. $\sqrt[3]{10}$

B. 2

C. $\sqrt{2} + \sqrt{5} - \sqrt{7}$

D. $\sqrt{6} - \sqrt{5} - \sqrt{11}$

E. $6x$

F. $4n$

Set 61

1A. 12

B. $6\sqrt{2}$

C. 10

D. 3

E. $3\sqrt{5}$

F. 17

2A. 6 feet, 8 feet, 10 feet

B. 4 cm, 7.5 cm, 8.5 cm

C. $\sqrt{105}$ ft $\approx$ 10.247 ft

D. 60 meters

E. 3.25 meters

F. $\sqrt{901}$ in $\approx$ 30.017 in

G. 3.25 centimeters

4. $72/\pi^2$ inches^2

Set 62

1A. $(x - 5)^2 - 28$

B. $(v + 2)^2 - 16$

C. $(x + 5/2)^2 + 7/4$

D. $(m - 7/2)^2 - 29/4$

E. $3(x - 2)^2 - 19$

F. $4(x + 1)^2 - 1$

G. $2(x + 3/4)^2 - 57/8$

H. $3(x - 5/2)^2 - 75/4$

I. $(x + 3)^2 - 5$

J. $(w - 1)^2 + 8$

K. $(x - 9/2)^2 - 85/4$

L. $(n + 1/2)^2 - 41/4$

M. $2(x + 3)^2 - 23$

N. $5(x - 1)^2 - 3$

O. $2(x + 5/2)^2 - 25/2$

P. $3(x - 5/6)^2 - 97/12$

2A. $\{-9, 1\}$

B. $\{1, 5\}$

C. $\{-5, 6\}$

D. $\{-11, 9\}$

E. $\{-5, 2\}$

F. $\{2, 4\}$

G. $\{1, 3/2\}$

H. $\{(-7 - \sqrt{61})/6, (-7 + \sqrt{61})/6\}$

I. $\{-1, -3/5\}$

J. no real solution

K. $\{-1, 13\}$

L. $\{-7, 3\}$

M. $\{-5, 2\}$

N. $\{-3, 13\}$

O. $\{(1 - \sqrt{37})/2, (1 + \sqrt{37})/2\}$

P. $\{-2, 0\}$

Q. $\{-5, -1/2\}$

R. $\{(9 - \sqrt{33})/8, (9 + \sqrt{33})/8\}$

S. $\{1/3, 3\}$

T. no real solution

Set 63

1A. $\{-5,-3\}$ B. $\{-11,1\}$ C. $\{(5-\sqrt{65})/2,(5+\sqrt{65})/2\}$

D. no real solution E. $\{2/3,5\}$ F. $\{(-6-\sqrt{11})/5,(-6+\sqrt{11})/5\}$

G. $\{1\}$ H. $\{-4,2\}$ I. $\{-14,-11/2\}$ J. $\{-9,13\}$

K. $\{2,7\}$ L. $\{(-7-\sqrt{61})/2,(-7+\sqrt{61})/2\}$ M. no real solution

N. $\{-1,3/2\}$ O. $\{(-2-\sqrt{5})/2,(-2+\sqrt{5})/2\}$ P. $\{-3\}$

Q. $\{(-ac-\sqrt{a^2c^2-4ab^2c})/(2bc),(-ac+\sqrt{a^2c^2-4ab^2c})/(2bc)\}$ R. $\{7,8\}$

3A. $x^2+7x-18=0$ B. $x^2-2x-4=0$ C. $x^2+16x+60=0$ D. $x^2-8x+9=0$

Set 64

1A. $\{-5,8\}$ B. $\{4,10\}$ C. $\{-5\}$ D. $\{-3,1\}$

E. $\{(-5-\sqrt{29})/2,(-5+\sqrt{29})/2\}$ F. $\{-\sqrt{7},\sqrt{7}\}$

G. $\{(-3-\sqrt{15})/2,(-3+\sqrt{15})/2\}$ H. $\{(11-\sqrt{41})/8,(11+\sqrt{41})/8\}$

I. $\{(-5-\sqrt{13})/2,(5-\sqrt{13})/2\}$ J. $\{(-3+\sqrt{37})/14,(3+\sqrt{37})/14\}$

K. $\{-13/4,5\}$ L. $\{-8,-3\}$ M. $\{(7-\sqrt{17})/4,(7+\sqrt{17})/4\}$

N. $\{-9,-7\}$ O. $\{-8,8\}$ P. $\{-6,7\}$ Q. $\{4,8\}$

R. $\{(7-\sqrt{33})/2,(7+\sqrt{33})/2\}$ S. $\{-5/2,0\}$

T. $\{(1-\sqrt{22})/3,(1+\sqrt{22})/3\}$ U. $\{(-5-\sqrt{10})/5,(-5+\sqrt{10})/5\}$

V. $\{(-7+\sqrt{89})/4,(7+\sqrt{89})/4\}$ W. $\{(-9-\sqrt{33})/6,(9-\sqrt{33})/6\}$

X. $\{2/3,4\}$ Y. $\{-4,9\}$ Z. $\{(-3-\sqrt{89})/4,(-3+\sqrt{89})/4\}$

2A. $\{-11,-1/5\}$ B. $\{-11/8,7/6\}$ C. $\{-8,5/2,26/3\}$ D. $\{-2,-5/3,-1,8\}$

E. $\{3/2\}$ F. $\{3/2\}$ G. $\{-11/6,1/2,9/2\}$ H. $\{-3,7/5,5\}$

Set 65

1A. -3 B. 5 C. 8 and 10 D. 9 and 11

E. 8 in by 13 in F. b: 6 ft; h: 8 ft G. 12 ft H. 20/3 yd

I. 12 in by 17 in J. 2 in K. 9 in L. 2 in

M. b: 13 cm; h: 7 cm N. 21 rows O. 6 units, 10 units P. 15 m

Q. 5/12 hr = 25 min R. 140 min = 2 hr 20 min

S. $(9-\sqrt{29})/2$ and $(9+\sqrt{29})/2$ T. $4-\sqrt{6}$ and $4+\sqrt{6}$

Set 66

1A. 6 B. -5 C. -32 D. -128 E. 1/9 F. 27

G. 121 H. 100 I. 19 J. -9 K. -4 L. 81

M. 512 N. 1/16 O. 49 P. 125 Q. 36 R. 64

2A. $49x^3y^{2/5}$ B. $-36m^{5/7}$ C. $8/v^{11/10}$ D. $3/x^{1/15}$ E. $-2b^{9/4}/5$ F. $8y^{3/5}/x^{21/2}$

G. $1/(32c^{5/12})$ H. $1/m^{25/6}$ I. $x^{7/6}y^{72/7}$ J. $27x^{3/4}y^2$ K. $100x^{5/3}/y^9$ L. $-24n^8$

M. $18w^{73/40}$ N. $-x^{11/3}/5$ O. $-54x^{11/20}$ P. $27/x^{63/10}$ Q. $n^{11/6}$ R. $x^{56/15}/y^{1/15}$

3A. $\sqrt[4]{x}$ B. $\sqrt{3}$ C. $x\sqrt[4]{x}$ D. $\sqrt[15]{m^{14}}$ E. $3\sqrt[28]{3}$ F. $\sqrt[5]{x}$

G. $\sqrt{2}$ H. $x^{24}\sqrt{x^7}$ I. $\sqrt[7]{n}$ J. $\sqrt[15]{2}$ 4A. $\{-32,32\}$ B. $\{4\}$

C. $\{-27,27\}$ D. $\{81\}$ 5A. 1/2 B. -5

Set 67

1A. $(x^{1/5}+2)(x^{1/5}+3)$ G. $(x^{1/3}-7)(x^{1/3}+3)$

B. $(v^{-1/2}+6)(v^{-1/2}-4)$ H. $(w^{-11/4}-5)^2$

C. $m^{1/2}(m-3)(m+3)$ I. $n^{1/2}(n^{1/2}-4)(n^{1/2}+4)$

D. $x^{1/3}(x^{1/3}+3)(x^{2/3}-3x^{1/3}+9)$ J. $x^{6/5}(x^{1/5}-2)(x^{2/5}+2x^{1/5}+4)$

E. $(6x+13)(2x+9)$ K. $3(10x-13)(5x-4)$

F. $5(x-13)(x^2-14x+61)$ L. $4(5x+1)(5x+2)(10x+3)$

2A. $\{1\}$ B. $\{0,49\}$ C. $\{1/5,1/2\}$ D. $\{-9,3\}$

E. $\{-2\sqrt{2},2\sqrt{2}\}$ F. $\{\sqrt[3]{100}/10,\sqrt[3]{9}/3\}$ G. $\{0,16\}$ H. $\{0,4\}$

I. $\{-\sqrt[3]{9}/3,-\sqrt[3]{25}/5\}$ J. $\{-1/2,7/2\}$ K. $\{-\sqrt[3]{9},\sqrt[3]{5}\}$ L. $\{-1/9,1/9\}$

3. x^4

Set 68

1.

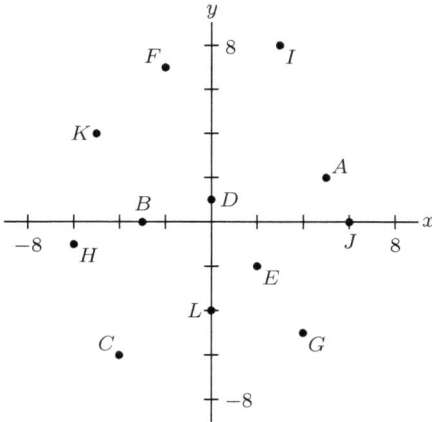

2A. II   E. I
   B. IV   F. II
   C. I   G. IV
   D. III   H. III

Set 69

1A. d: 10; m: $(7, 5)$
 B. d: 5; m: $(-25/2, -5)$
 C. d: 1; m: $(-3, 17/2)$
 D. d: $\sqrt{13}$; m: $(0, 17/2)$
 E. d: $2\sqrt{13}$; m: $(-6, 2)$
 F. d: $\sqrt{a^2 + b^2}$; m: $(-a/2, b/2)$
 G. d: $\sqrt{17}$; m: $(-1/2, -11/6)$
 H. d: $\sqrt{64621}/210$; m: $(-11/42, 41/20)$
2A. 36   B. 28

 I. d: 13; m: $(1, 11/2)$
 J. d: 17; m: $(-15/2, 0)$
 K. d: 6; m: $(1, 5)$
 L. d: $\sqrt{34}$; m: $(13/2, 11/2)$
 M. d: $3\sqrt{10}$; m: $(-27/2, -13/2)$
 N. d: $\sqrt{9a^2 + 49b^2}$; m: $(3a/2, 7b/2)$
 O. d: $\sqrt{89}/5$; m: $(-2/5, 23/10)$
 P. d: $\sqrt{122}/6$; m: $(17/12, 41/12)$
 C. 33   **3.** B

Set 70

1A. $y = (2x - 10)/5$ B. $x = (-3 + \sqrt{y})^2$ C. $y = \pm\sqrt{25 - x^2}$
 D. $x = -4y - 8$ E. $y = (8 - 4x)^3/5$ F. $x = 2 \pm\sqrt{y + 9}$

2A.

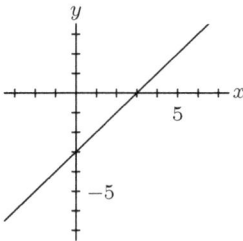

B.

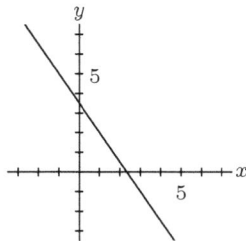

C.

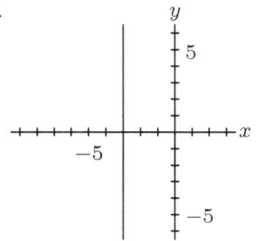

D.

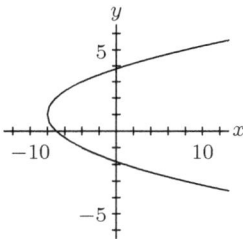

E.

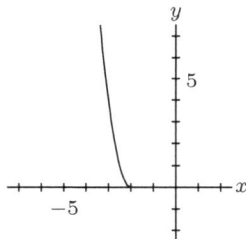

F.

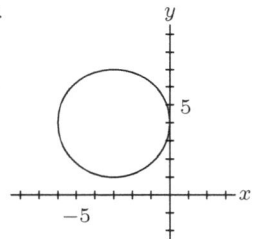

G.

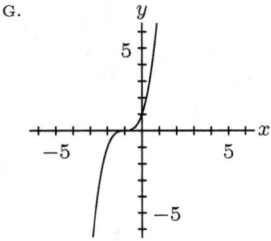

H.

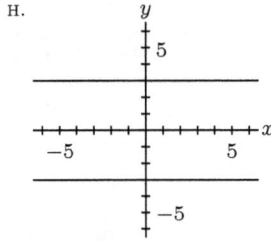

I.

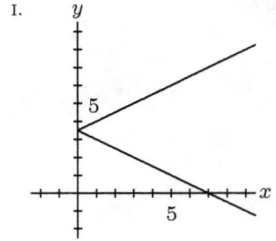

J.

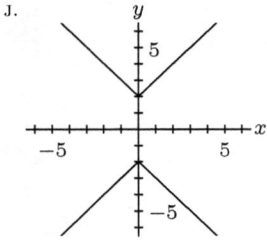

K.

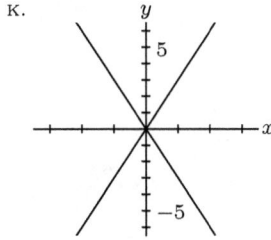

L.

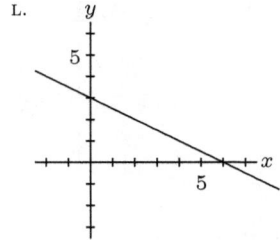

M.

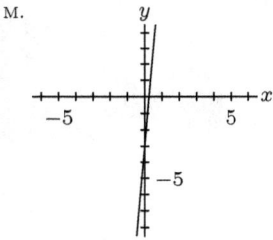

N.

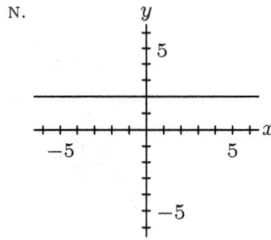

O.

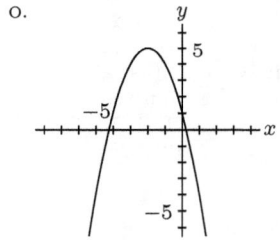

P.

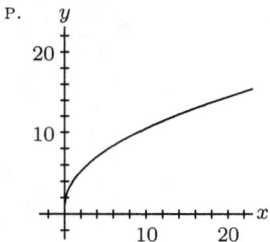

Q.

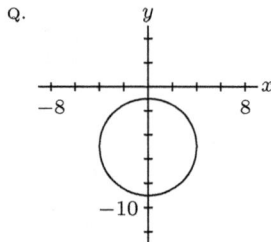

R.

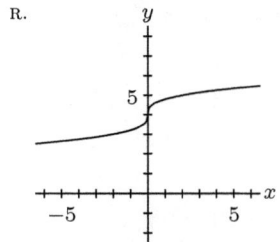

S.

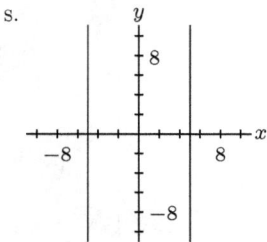

T.

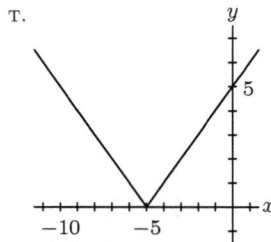

U.

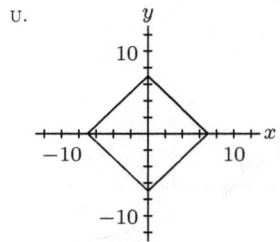

3A. x-int: $5/2$; y-int: 3

D. x-int: -8; y-int: 12

B. x-int: -2; y-int: 8

E. x-ints: -3, 7; y-int: none

C. x-ints: -8, 8; y-ints: -8, 8

F. x-ints: $-\sqrt{7}$, $\sqrt{7}$; y-int: -7

Set 71
1A. 4   B. $-1/3$ C. 0   D. -6 E. $-2/9$ F. undefined
 2. -3 **3**. $-5/2$ **4**. $22\sqrt{2}$ **5**. 6
6A. zero   B. negative   C. positive   D. undefined
7A. $(-2,-2)$, $(2,8)$, $(4,13)$ D. $(-5,9)$, $(1,5)$, $(4,3)$
  B. $(-10,6)$, $(-9,-1)$, $(-7,-15)$ E. $(0,1)$, $(1,7)$, $(2,13)$
  C. $(0,-4)$, $(8,-4)$, $(20,-4)$ F. $(6,0)$, $(6,5)$, $(6,10)$
8A. pts: $(0,5)$, $(3,13)$; $m = 8/3$ E. pts: $(0,-7)$, $(4,-9)$; $m = -1/2$
  B. pts: $(-8,0)$, $(-3,-1)$; $m = -1/5$ F. pts: $(11,0)$, $(18,6)$; $m = 6/7$
  C. pts: $(0,-9/8)$, $(9/2,0)$; $m = 1/4$ G. pts: $(-1/5,0)$, $(0,-1/4)$; $m = -5/4$
  D. pts: $(1,-9)$, $(2,-9)$; $m = 0$ H. pts: $(12,0)$, $(12,10)$; m is undefined

Set 72
1A. $y = 4x - 3$; $4x - y = 3$ D. $y = -3x + 5$; $3x + y = 5$
  B. $y = (-5/3)x$; $5x + 3y = 0$ E. $y = (2/7)x + 2$; $2x - 7y = -14$
  C. $y = (9/2)x + 1/11$; $99x - 22y = -2$ F. $y = (-8/5)x - 3/4$; $32x + 20y = -15$
2A. $y = (5/6)x + 13/2$; $5x - 6y = -39$ D. $y = (3/2)x + 7/2$; $3x - 2y = -7$
  B. $y = -8x - 5$; $8x + y = -5$ E. $y = (-1/4)x - 17/4$; $x + 4y = -17$
  C. $y = 3$; $y = 3$ F. no slope-int form; $x = 1$
3A. $y = (1/2)x + 3/2$; $x - 2y = -3$ D. $y = (-2/5)x - 6/5$; $2x + 5y = -6$
  B. $y = -3x - 19$; $3x + y = -19$ E. $y = (17/2)x - 37$; $17x - 2y = 74$
  C. no slope-int form; $x = -7$ F. $y = 7$; $y = 7$
4A. x-int: 10; y-int: 2   B. x-int: 2/3; y-int: $-2/7$ C. x-int: 3/8; y-int: -1

D. x-int: 4; y-int: none   E. x-int: -3; y-int: 4   F. x-int: $-3/8$; y-int: -3

G. x-int: 8/3; y-int: 2   H. x-int: none; y-int: -3 **5**. $F = (9/5)C + 32$

6A. eqn: $y = -5x + 1$; $m = -5$; $b = 1$
  B. eqn: $y = (1/3)x - 7$; $m = 1/3$; $b = -7$
  C. eqn: $y = 88x + 0$; $m = 88$; $b = 0$
  D. eqn: $y = (7/2)x + 3/2$; $m = 7/2$; $b = 3/2$
  E. eqn: $y = -\frac{5}{36}x - \frac{20}{9}$; $m = -\frac{5}{36}$; $b = -\frac{20}{9}$
  F. eqn: $y = (2/5)x - 20$; $m = 2/5$; $b = -20$

  G. eqn: $y = 2x - 11$; $m = 2$; $b = -11$
  H. eqn: $y = (-1)x + 6$; $m = -1$; $b = 6$
  I. eqn: $y = 0x - 23$; $m = 0$; $b = -23$
  J. eqn: $y = (-1/5)x + 0$; $m = -1/5$; $b = 0$
  K. eqn: $y = (14/3)x + 14$; $m = 14/3$; $b = 14$
  L. eqn: $y = -\frac{33}{16}x + \frac{11}{2}$; $m = -\frac{33}{16}$; $b = \frac{11}{2}$

Set 73

1A. $m_{\parallel} = -2/9$; $m_{\perp} = 9/2$ B. $m_{\parallel} = 5$; $m_{\perp} = -1/5$ C. $m_{\parallel} = 0$; $m_{\perp}$ is undefined
  D. $m_{\parallel} = 1/4$; $m_{\perp} = -4$ E. $m_{\parallel} = -8$; $m_{\perp} = 1/8$ F. $m_{\parallel}$ is undefined; $m_{\perp} = 0$
2A. $x = -4$ B. $y = 6$ C. $y = 13$ D. $y = -5$ E. $y = -1$ F. $x = -7$
  G. $x = 9$ H. $x = -3$ **3**A. $y = 5x + 6$ B. $y = -3x - 29$ C. $y = -3$
  D. $y = (2/9)x - 25/9$ E. $y = (8/5)x - 7$ F. $y = (2/3)x - 10/3$
  G. $y = 15$ H. $y = -8x - 73$ **5.** $20x + 6y = 101$

Set 74

1A. $(4, -2)$ B. no solution   C. $(0, 3)$ D. no solution
2A. solution: $(-2, 5)$ B. solution: $(6, -4)$ C. solution: $(-6, -1)$

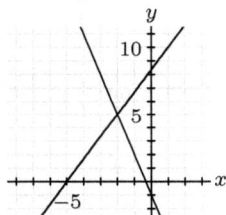
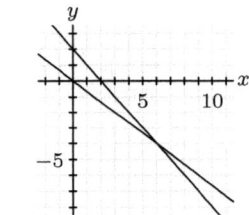
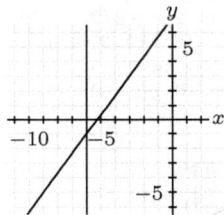

D. solution: $(-3, -2)$ E. solution: $(1, 2)$ F. solution: $(4, 3)$

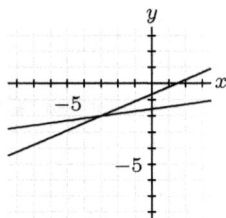
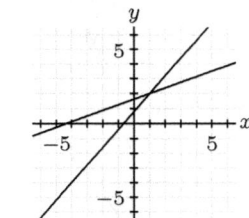
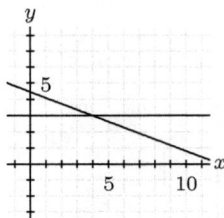

Set 75

1A. 7 to first; -2 to second   B. 3 to first; -5 to second   **2**A. 2 to first; 3 to second
  B. -1 to first; 4 to second   **3**A. $\{(-2, 5)\}$ B. $\{(9, -2)\}$ C. $\{(3, -5)\}$ D. $\{(7/15, 1/3)\}$
  E. $\emptyset$ F. $\{(-5/3, -1/2)\}$ G. $\{(2, 5)\}$ H. $\{(13, 8)\}$ I. $\{(-12, 20)\}$
  J. $\{(-3, -1)\}$ K. $\{(-17/11, 35/11)\}$ L. $\{(x, y) \in \mathbf{R}^2 \mid y = (6x + 9)/7\}$
  M. $\{(-2, 9)\}$ N. $\{(6, -4)\}$ O. $\{(3, 5/11)\}$ P. $\{(-3, -5)\}$ Q. $\{(56, 6)\}$ R. $\{(-6, -10)\}$

Set 76

1A. $\{(7, 2)\}$ B. $\{(5, -6)\}$ C. $\{(-1, -3)\}$ D. $\{(-2, 3)\}$ E. $\{(x, y) \in \mathbf{R}^2 \mid y = 6x + 4\}$
  F. $\{(-5, -8)\}$ G. $\{(3, 1)\}$ H. $\{(-4, -7)\}$ I. $\{(2, -4)\}$ J. $\{(6, -10)\}$ K. $\emptyset$
  L. $\{(-9, 16)\}$ **2**A. $\{(-2, -2), (1, 1)\}$ B. $\{(-2, -6), (1, 3), (4, -6)\}$
  C. $\{(4, -3), (-3, 4)\}$ D. $\{(2, 6), (-5, -15)\}$ E. $(-5, 8)$
  F. $\{(3, -1), (6, 1)\}$ G. $\{(-2, -2), (-3, -1)\}$ H. $\{(0, -3), (24/13, -15/13)\}$
  I. $\{(-1, 3), (125/8, -13/2)\}$ J. $(-1, 3)$ **3**A. 15/2 and 31/2   B. 37
  C. O: 78; P: 41   D. d: 14; q: 23   E. y: 4 g; b: 8 g   F. a: \$19; c: \$11   G. R: \$3.35; S: \$8.60
  H. $41°$ and $49°$ I. 4\%: \$600; 5\%: \$200   J. 90 in^3 **4.** 5   **5.** 7

Set 77

1A.

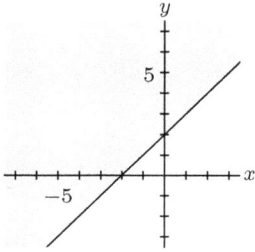

B.

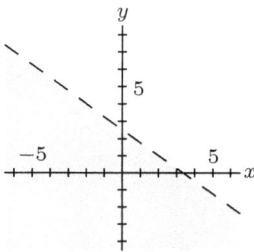

C.

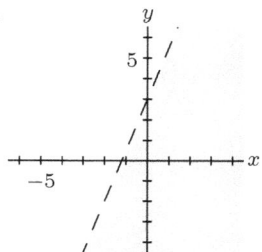

D.

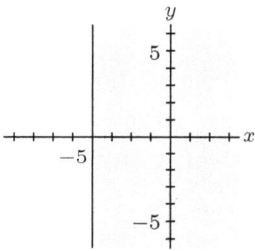

E.

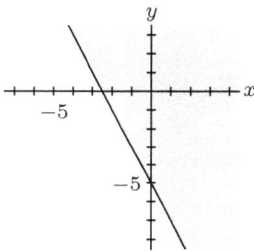

F.

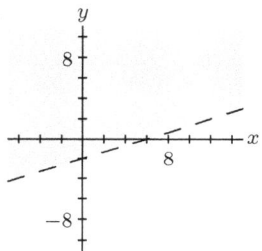

G.

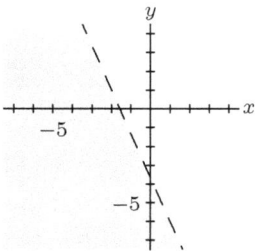

H.

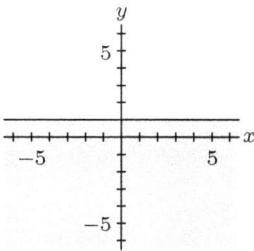

2A.

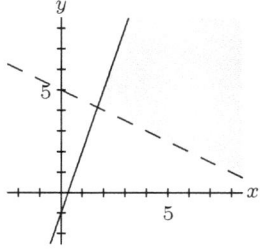

B.

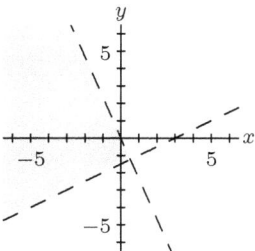

C.

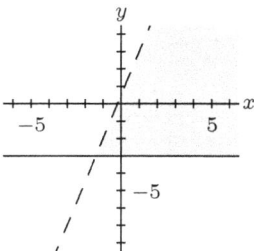

D.

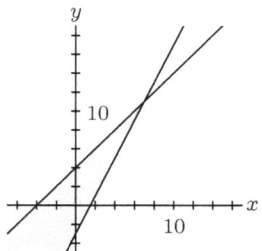

E.

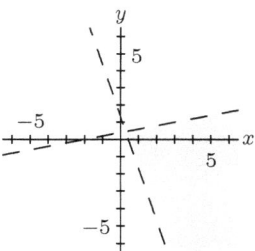

F.

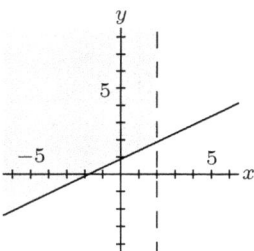

3A.

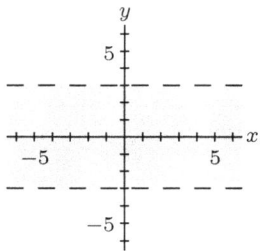

B.

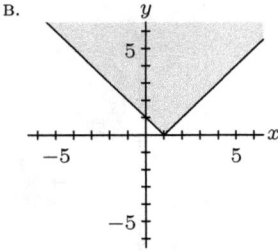

C.

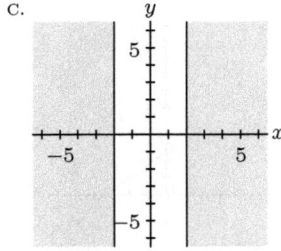

D.

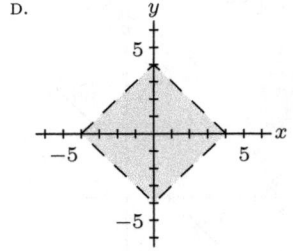

Set 78

1A. $\{(3,-1,-5)\}$ B. $\{(-4,8,-3)\}$ C. $\{(5,10,-4)\}$ D. $\{(x,y,z) \in \mathbf{R}^3 \mid z = 7 - 4x + 2y\}$

E. $\emptyset$ F. $\{(1,4,7)\}$ G. $\{(0,-1,-9)\}$ H. $\{(-6,6,2)\}$ I. $\emptyset$

J. $\{(x,y,z) \in \mathbf{R}^3 \mid x = 68 - 12z,\ y = 5z - 29\}$ **2**A. 683 B. 7, 18, 31

C. A: $115°$, B: $23°$, C: $42°$ D. 10%: 15 gal; 20%: 5 gal; 40%: 30 gal E. $\{-4,5,11\}$

Set 79

2A. $(1,-4)$ satisfies the eqn., but $(-1,-4)$ does not. B. $(7,1)$ satisfies the eqn., but $(7,-1)$ does not. C. $(6,-2)$ satisfies the eqn., but $(-6,2)$ does not.

3A. symmetric w.r.t. origin B. symmetric w.r.t. y-axis C. symmetric w.r.t. x-axis

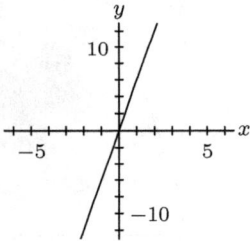

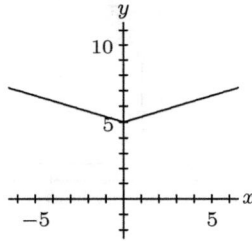

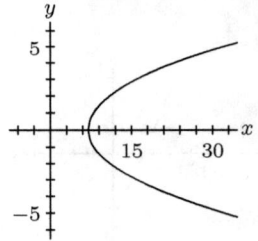

D. none

E. symmetric w.r.t. x-axis

F. symmetric w.r.t. x-axis, y-axis, and origin

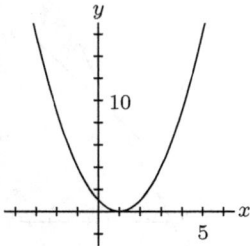

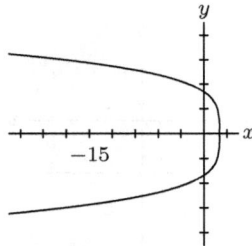

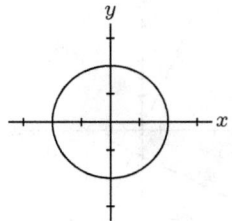

G. none

H. symmetric w.r.t. y-axis

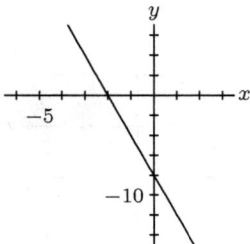

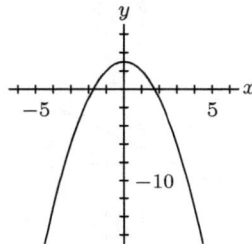

Set 80

1A. 1   B. -7 C. $2e - 7$ D. $2n + 3$ E. -13 F. $-11/2$ G. $10m - 7$
H. $2x^2 - 2x - 5$ 2A. 2   B. $-1/4$ C. $4y^2 + 10y + 2$ D. $v - 6 + 5\sqrt{v - 8}$
E. 152   F. $36z^2 + 18z - 2$ G. $w^6 - 4w^5 + 12w^4 - 21w^3 + 26w^2 - 20w + 2$
H. $\pi^2 + 5\pi + 2$ 3A. 5   B. 3   C. 5   D. -37 4A. 4   B. 42
C. 19   D. 0   5A. 7   B. 7   C. 242   D. -26 E. 50   F. 7
6A. 35   B. undefined   C. $7\sqrt{17}$ D. 21   E. 25   F. undefined
7A. 8   B. $2a + h + 4$ C. $-4/[(a + h - 1)(a - 1)]$ D. $10a + 5h$
E. $3a^2 + 3ah + h^2$ F. $3/[\sqrt{3a + 3h} + \sqrt{3a}]$ 8. x

Set 81

1A. d: $\{-5, -1, 0, 2, 3\}$; r: $\{-3, 2, 7, 8, 10\}$; yes, it is a function
B. d: $\{0, 1, 2, 4, 5, 7, 12\}$; r: $\{5, 6, 7, 8, 9, 11, 13, 16\}$; no, it is not a function
C. d: $\{6\}$; r: $\{-1, 2, 10\}$; no, it is not a function
D. d: $\{-9, -4, 3, 5, 11\}$; r: $\{5\}$; yes, it is a function

Set 82

1A. $\mathbf{R}$ B. $\{x \in \mathbf{R} \mid x \neq 3\}$ C. $[-3/10, \infty)$ D. $\mathbf{R}$
E. $\{x \in \mathbf{R} \mid x \neq -7/4 \text{ AND } x \neq 4\}$ F. $(2, \infty)$ G. $(-\infty, -6] \cup [3, \infty)$
H. $[2, 7) \cup (7, \infty)$ I. $[-5, -3] \cup [3, \infty)$ J. $\{x \in \mathbf{R} \mid x \neq -9 \text{ AND } x \neq 5\}$
K. $\mathbf{R}$ L. $(-\infty, 2]$ M. $\{x \in \mathbf{R} \mid x \neq -2 \text{ AND } x \neq 4\}$
N. $\mathbf{R}$ O. $(-3/2, \infty)$ P. $(-\infty, -3) \cup (-5/2, \infty)$
Q. $[17, \infty)$ R. $(-\infty, -2] \cup (-1, 2) \cup (2, \infty)$ 2A. $\mathbf{R}$
B. $\{y \in \mathbf{R} \mid y \neq 0\}$ C. $\{y \in \mathbf{R} \mid y \neq 5/2\}$ D. $\mathbf{R}$ E. $[0, \infty)$
F. $(-\infty, 12]$ G. $[-22, 3]$ H. $\mathbf{R}$ I. $\{y \in \mathbf{R} \mid y \neq 13\}$
J. $\{7/2\}$ K. $\mathbf{R}$ L. $[0, \infty)$ M. $[-6, \infty)$ N. $[10, 26]$

Set 83

1A. $f(x) = (5x - 15)/3$ B. $f(x) = 1/(4x)$ C. $f(x) = 9x/(1 - x)$
D. $f(x) = (5 - 8x^2)/4$ E. $f(x) = 4x/7$ F. $f(x) = 7/(11 - 2x)$
2A. $V(r) = (4/3)\pi r^3$ B. $C(r) = 2\pi r$ C. $A(b) = 4b$ D. $A(h) = 5h/2$
E. $d(t) = 60t$ F. $I(t) = 11.25t$ G. $l(w) = 100/w$ H. $S = 6V^{2/3}$
I. $w(l) = (108 - 6l)/(l + 6)$ J. $A(C) = C^2/(4\pi)$ 3. $A(w) = 5\pi w^2/2$

Set 84

1A. yes   B. yes   C. no   D. yes   E. yes   F. no   G. yes   H. no

2A.   B.   C.

D.   E.   F.

G.

H.

I.

J.

K.

L.

M.

N.

O.

P.

Q.

Set 85

1A. turning pts.: $(-3, 3)$, $(-1, -6)$, and $(2, 1)$;
inc. on $[-5, -3]$ and $[-1, 2]$;
dec. on $[-3, -1]$ and $[2, 4]$;
abs. max. of 3 at $x = -3$;
abs. min. of -6 at $x = -1$;
rel. max. of 1 at $x = 2$;
rel. min. of -4 at $x = -5$;
rel. min. of -2 at $x = 4$.

B. turning pts.: $(1, -2)$ and $(6, 4)$;
inc. on $[1, 6]$;
dec. on $(-\infty, 1]$ and $[6, \infty)$;
no abs. max.;
no abs. min.;
rel. max. of 4 at $x = 6$;
rel. min. of -2 at $x = 1$.

C. turning pts.: $(-3, 2)$, $(-1, 6)$, $(1, 2)$, and $(3, 6)$;
 inc. on $[-3, -1]$ and $[1, 3]$;
 dec. on $[-4, -3]$, $[-1, 1]$, and $[3, 5]$;
 abs. max. of 6 at $x = -1$ and $x = 3$;
 abs. min. of -2 at $x = 5$;
 rel. max. of 4 at $x = -4$;
 rel. min. of 2 at $x = -3$ and $x = 1$.

D. turning pt.: $(-2, 5)$;
 inc. on $(-\infty, -2]$;
 dec. on $[-2, \infty)$;
 abs. max. of 5 at $x = -2$;
 no abs. min.

2A. 6 B. -30 C. $5/7$ D. 31 E. $1/7$
 F. -1 G. 22 H. $-21/77$ I. 76 J. 0

Set 86

1A. $(1, 10)$, $(7, -1)$, $(10, 0)$
 B. $(-6, 13)$, $(0, 2)$, $(3, 3)$
 C. $(-6, -50)$, $(0, 5)$, $(3, 0)$
 D. $(-36, 10)$, $(0, -1)$, $(18, 0)$
 E. $(-10, 1)$, $(-4, -10)$, $(-1, -9)$
 F. $(-3, 2)$, $(0, -1/5)$, $(3/2, 0)$
 G. $(1/2, 29)$, $(-1, -15)$, $(-7/4, -11)$

 H. $(-6, 8)$, $(0, -3)$, $(3, -2)$
 I. $(-16, 10)$, $(-10, -1)$, $(-7, 0)$
 J. $(2, 10)$, $(0, -1)$, $(-1, 0)$
 K. $(-6, 5)$, $(0, -1/2)$, $(3, 0)$
 L. $(2, 17)$, $(8, 6)$, $(11, 7)$
 M. $(-24, 90)$, $(0, -9)$, $(12, 0)$
 N. $(2/3, -6)$, $(5/3, 5)$, $(13/6, 4)$

2A. $(-6, 1)$, $(-5, 0)$, $(-2, 3)$ B. $(-1, 9/2)$, $(0, 4)$, $(3, 11/2)$ C. $\left(-\frac{8}{3}, -3\right)$, $(-3, -1)$, $(-4, -7)$

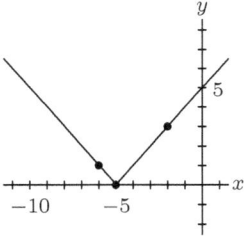

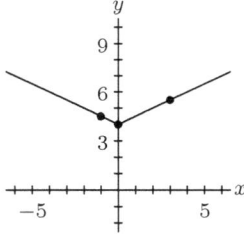

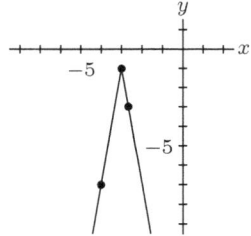

3A. $(-2, 16)$, $(-1, 2)$, $(0, 0)$, $(1, -2)$, $(2, -16)$ B. $(-6, -15)$, $(-3, -8)$, $(0, -7)$, $(3, -6)$, $(6, 1)$ C. $(9/2, 10)$, $(19/4, 3)$, $(5, 2)$, $(21/4, 1)$, $(11/2, -6)$

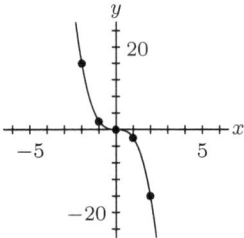

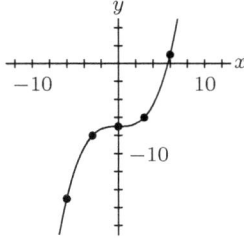

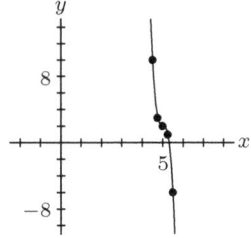

4A. $(-4, 3)$, $(-3, 1)$, $(-1, 5)$, $(1, 1)$, $(3, 5)$, $(5, -3)$ B. $(2, 4)$, $(5/2, 2)$, $(7/2, 6)$, $(9/2, 2)$, $(11/2, 6)$, $(13/2, -2)$ C. $(11, 19)$, $(10, 11)$, $(8, 27)$, $(6, 11)$, $(4, 27)$, $(2, -5)$

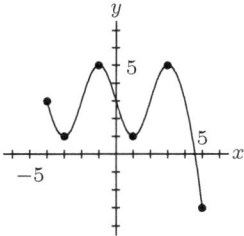

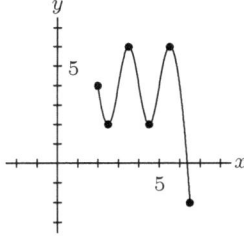

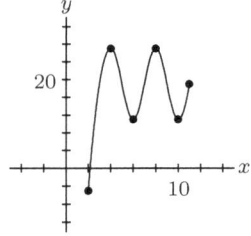

5A. $(5/4, -4)$, $(3/4, 3)$, $(1/4, -6)$, $(-1/2, 1)$, $(-1, -2)$

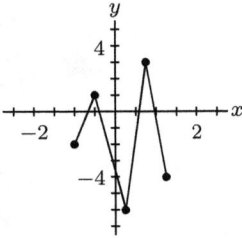

B. $(-14, -12)$, $(-12, 9)$, $(-10, -18)$, $(-7, 3)$, $(-5, -6)$

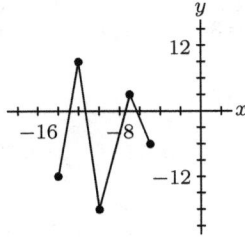

C. $(-3/2, -42/5)$, $(1/2, -43/5)$, $(-1/2, -77/10)$, $(2, -79/10)$, $(3, -41/5)$

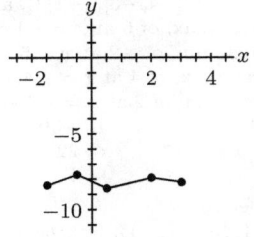

Set 87

1A. $(-3, 11)$, $(0, 5)$, $(-2, 1)$

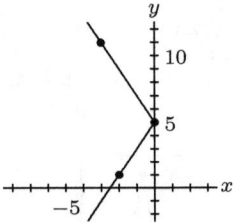

B. $(-6, 6)$, $(-3, 0)$, $(-5, -4)$

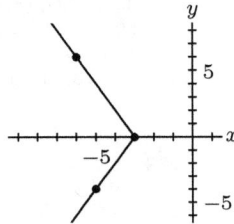

C. $(4, 4)$, $(7, -2)$, $(5, -6)$

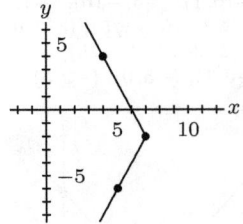

2A. $(3, -4)$, $(0, -1)$, $(-3, -4)$, $(0, -7)$

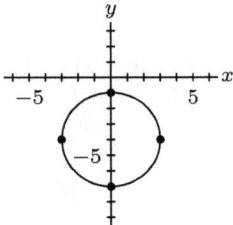

B. $(4, 9)$, $(1, 12)$, $(-2, 9)$, $(1, 6)$

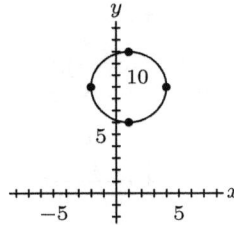

C. $(1, 5)$, $(-2, 8)$, $(-5, 5)$, $(-2, 2)$

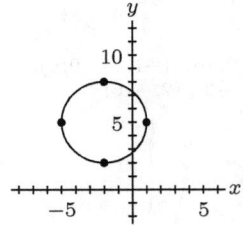

3A. $(5, 0)$, $(3, 5)$, $(1, 0)$, $(3, -5)$

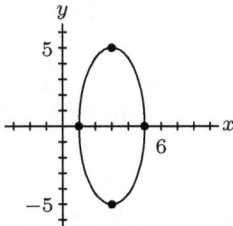

B. $(2, 10)$, $(0, 15)$, $(-2, 10)$, $(0, 5)$

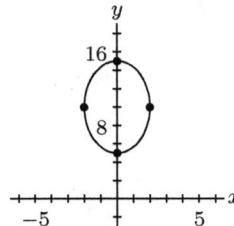

C. $(0, -8)$, $(-2, -3)$, $(-4, -8)$, $(-2, -13)$

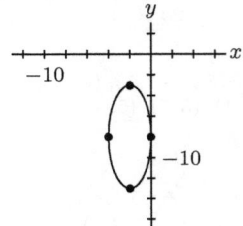

Set 88

1A. even	B. odd	C. neither	D. even	E. even	F. neither
G. odd	H. odd	I. even	J. neither	K. neither	L. neither
M. even	N. even	O. even	P. odd	**3**A. 13	B. −17

Set 89

1A. no   B. yes   C. no   D. no   E. yes   F. no   G. no   H. yes

2A. no   B. yes   C. no   D. yes   E. yes   F. no   G. yes   H. no

Set 90

1A. 1   B. 3   C. 1   D. undefined   E. 5   F. 7   G. 6

H. -2 I. 6/7   J. 1   K. 42   L. 1   M. 6   N. 3

2A. $(f + g)(x) = 3x + 15$, $(f - g)(x) = 5x - 1$, $(fg)(x) = -4x^2 + 25x + 56$,
$(f/g)(x) = (4x + 7)/(8 - x)$, $(f \circ g)(x) = 39 - 4x$, $(g \circ f)(x) = 1 - 4x$;
$\mathbb{D}_{f+g} = \mathbb{D}_{f-g} = \mathbb{D}_{fg} = \mathbb{D}_{f \circ g} = \mathbb{D}_{g \circ f} = \mathbf{R}$, $\mathbb{D}_{f/g} = \{x \in \mathbf{R} \mid x \neq 8\}$; $(f + g)(-3) = 6$,
$(f - g)(6) = 29$, $(fg)(2) = 90$, $(f/g)(5) = 9$, $(f \circ g)(-2) = 47$, $(g \circ f)(3) = -11$

B. $(f + g)(x) = (2x^3 - 17x^2 + 20x + 76)/(2x^2 - 7x - 15)$,
$(f - g)(x) = (2x^3 - 17x^2 + 20x + 74)/(2x^2 - 7x - 15)$, $(fg)(x) = 1/(2x + 3)$,
$(f/g)(x) = 2x^3 - 17x^2 + 20x + 75$, $(f \circ g)(x) = -(10x^2 - 35x - 76)/(2x^2 - 7x - 15)$,
$(g \circ f)(x) = 1/(2x^2 - 27x + 70)$; $\mathbb{D}_{g \circ f} = \{x \in \mathbf{R} \mid x \neq 7/2$ AND $x \neq 10\}$,
$\mathbb{D}_{f+g} = \mathbb{D}_{f-g} = \mathbb{D}_{fg} = \mathbb{D}_{f/g} = \mathbb{D}_{f \circ g} = \{x \in \mathbf{R} \mid x \neq -3/2$ AND $x \neq 5\}$;
$(f + g)(-3) = -191/24$, $(f - g)(6) = 14/15$, $(fg)(2) = 1/7$, $(f/g)(5)$ is undefined,
$(f \circ g)(-2) = -34/7$, $(g \circ f)(3) = 1/7$

C. $(f + g)(x) = |7x + 1| - 6$, $(f - g)(x) = |7x + 1| + 6$, $(fg)(x) = -6|7x + 1|$,
$(f/g)(x) = -|7x + 1|/6$, $(f \circ g)(x) = 41$, $(g \circ f)(x) = -6$;
$\mathbb{D}_{f+g} = \mathbb{D}_{f-g} = \mathbb{D}_{fg} = \mathbb{D}_{f/g} = \mathbb{D}_{f \circ g} = \mathbb{D}_{g \circ f} = \mathbf{R}$, $(f + g)(-3) = 14$,
$(f - g)(6) = 49$, $(fg)(2) = -90$, $(f/g)(5) = -6$, $(f \circ g)(-2) = 41$, $(g \circ f)(3) = -6$

D. $(f + g)(x) = (4x - 5)/(x + 2)$, $(f - g)(x) = (2x - 5)/(x + 2)$,
$(fg)(x) = (3x^2 - 5x)/(x + 2)^2$, $(f/g)(x) = (3x - 5)/x$, $(f \circ g)(x) = -(2x + 10)/(3x + 4)$,
$(g \circ f)(x) = (3x - 5)/(5x - 1)$; $\mathbb{D}_{f+g} = \mathbb{D}_{f-g} = \mathbb{D}_{fg} = \{x \in \mathbf{R} \mid x \neq -2\}$,
$\mathbb{D}_{f/g} = \{x \in \mathbf{R} \mid x \neq -2$ AND $x \neq 0\}$, $\mathbb{D}_{f \circ g} = \{x \in \mathbf{R} \mid x \neq -2$ AND $x \neq -4/3\}$,
$\mathbb{D}_{g \circ f} = \{x \in \mathbf{R} \mid x \neq -2$ AND $x \neq 1/5\}$; $(f + g)(-3) = 17$, $(f - g)(6) = 7/8$,
$(fg)(2) = 1/8$, $(f/g)(5) = 2$, $(f \circ g)(-2)$ is undefined, $(g \circ f)(3) = 2/7$

E. $(f + g)(x) = x^2 + 2x - 3$, $(f - g)(x) = x^2 - 9$, $(fg)(x) = x^3 + 4x^2 - 3x - 18$,
$(f/g)(x) = x - 2$, $(f \circ g)(x) = x^2 + 7x + 6$, $(g \circ f)(x) = x^2 + x - 3$;
$\mathbb{D}_{f+g} = \mathbb{D}_{f-g} = \mathbb{D}_{fg} = \mathbb{D}_{f \circ g} = \mathbb{D}_{g \circ f} = \mathbf{R}$, $\mathbb{D}_{f/g} = \{x \in \mathbf{R} \mid x \neq -3\}$, $(f + g)(-3) = 0$,
$(f - g)(6) = 27$, $(fg)(2) = 0$, $(f/g)(5) = 3$, $(f \circ g)(-2) = -4$, $(g \circ f)(3) = 9$

F. $(f + g)(x) = x^2 - 4x - 5 + \sqrt{x + 9}$, $(f - g)(x) = -x^2 + 4x + 5 + \sqrt{x + 9}$,
$(fg)(x) = (x^2 - 4x - 5)\sqrt{x + 9}$, $(f/g)(x) = (\sqrt{x + 9})/(x^2 - 4x - 5)$, $(f \circ g)(x) = |x - 2|$,
$(g \circ f)(x) = x + 4 - 4\sqrt{x + 9}$; $\mathbb{D}_{f+g} = \mathbb{D}_{f-g} = \mathbb{D}_{fg} = \mathbb{D}_{g \circ f} = [-9, \infty)$, $\mathbb{D}_{f \circ g} = \mathbf{R}$,
$\mathbb{D}_{f/g} = [-9, -1) \cup (-1, 5) \cup (5, \infty)$, $(f + g)(-3) = 16 + \sqrt{6}$, $(f - g)(6) = -7 + \sqrt{15}$,
$(fg)(2) = -9\sqrt{11}$, $(f/g)(5)$ is undefined, $(f \circ g)(-2) = 4$, $(g \circ f)(3) = 7 - 8\sqrt{3}$

G. $(f + g)(x) = 15/(2x)$, $(f - g)(x) = 9/(2x)$, $(fg)(x) = 9/x^2$, $(f/g)(x) = 4$, $(f \circ g)(x) = 4x$,
$(g \circ f)(x) = x/4$; $\mathbb{D}_{f+g} = \mathbb{D}_{f-g} = \mathbb{D}_{fg} = \mathbb{D}_{f/g} = \mathbb{D}_{f \circ g} = \mathbb{D}_{g \circ f} = \{x \in \mathbf{R} \mid x \neq 0\}$,
$(f + g)(-3) = -5/2$, $(f - g)(6) = 3/4$, $(fg)(2) = 9/4$, $(f/g)(5) = 4$,
$(f \circ g)(-2) = -8$, $(g \circ f)(3) = 3/4$

H. $(f + g)(x) = (17x + 21)/4$, $(f - g)(x) = -(15x + 35)/4$, $(fg)(x) = (4x^2 - 21x - 49)/4$,
$(f/g)(x) = (x - 7)/(16x + 28)$, $(f \circ g)(x) = x$, $(g \circ f)(x) = x$;
$\mathbb{D}_{f+g} = \mathbb{D}_{f-g} = \mathbb{D}_{fg} = \mathbb{D}_{f \circ g} = \mathbb{D}_{g \circ f} = \mathbf{R}$, $\mathbb{D}_{f/g} = \{x \in \mathbf{R} \mid x \neq -7/4\}$,
$(f + g)(-3) = -15/2$, $(f - g)(6) = -125/4$, $(fg)(2) = -75/4$, $(f/g)(5) = -1/54$,
$(f \circ g)(-2) = -2$, $(g \circ f)(3) = 3$

5A. $(f \circ f)(x) = 25x$, $\mathbb{D}_{f \circ f} = \mathbf{R}$, $(f \circ f)(4) = 100$, $(f \circ f)(0) = 0$, $(f \circ f)(-3) = -75$

B. $(f \circ f)(x) = x/4$, $\mathbb{D}_{f \circ f} = \mathbf{R}$, $(f \circ f)(4) = 1$, $(f \circ f)(0) = 0$, $(f \circ f)(-3) = -3/4$

C. $(f \circ f)(x) = \sqrt[4]{x}$, $\mathbb{D}_{f \circ f} = [0, \infty)$, $(f \circ f)(4) = \sqrt{2}$, $(f \circ f)(0) = 0$, $(f \circ f)(-3) \notin \mathbf{R}$

D. $(f \circ f)(x) = \sqrt{4 - 6\sqrt{4 - 6x}}$, $\mathbb{D}_{f \circ f} = [16/27, 2/3]$, $(f \circ f)(4) \notin \mathbf{R}$, $(f \circ f)(0) \notin \mathbf{R}$,
$(f \circ f)(-3) \notin \mathbf{R}$

E. $(f \circ f)(x) = (5 - 2x)/(5x - 37)$, $\mathbb{D}_{f \circ f} = \{x \in \mathbf{R} \mid x \neq 6$ AND $x \neq 37/5\}$,
$(f \circ f)(4) = 3/17$, $(f \circ f)(0) = -5/37$, $(f \circ f)(-3) = -11/52$

F. $(f \circ f)(x) = 64x - 27$, $\mathbb{D}_{f \circ f} = \mathbf{R}$, $(f \circ f)(4) = 229$, $(f \circ f)(0) = -27$, $(f \circ f)(-3) = -219$

G. $(f \circ f)(x) = x$, $\mathbb{D}_{f \circ f} = \{x \in \mathbf{R} \mid x \neq 0\}$, $(f \circ f)(4) = 4$, $(f \circ f)(0)$ is undefined,
$(f \circ f)(-3) = -3$

H. $(f \circ f)(x) = \sqrt{-7 + 3\sqrt{3x - 7}}$, $\mathbb{D}_{f \circ f} = [112/27, \infty)$, $(f \circ f)(4) \notin \mathbf{R}$, $(f \circ f)(0) \notin \mathbf{R}$,
$(f \circ f)(-3) \notin \mathbf{R}$

I. $(f \circ f)(x) = \sqrt[9]{x}$, $\mathbb{D}_{f \circ f} = \mathbf{R}$, $(f \circ f)(4) = \sqrt[9]{4}$, $(f \circ f)(0) = 0$, $(f \circ f)(-3) = -\sqrt[9]{3}$

J. $(f \circ f)(x) = -(5x + 20)/(2x + 3)$, $\mathbb{D}_{f \circ f} = \{x \in \mathbf{R} \mid x \neq -4 \text{ AND } x \neq -3/2\}$,
$(f \circ f)(4) = -40/11$, $(f \circ f)(0) = -20/3$, $(f \circ f)(-3) = 5/3$

6A. $f(x) = \sqrt[4]{x}$, $g(x) = 11x + 6$

  B. $f(x) = x^2 + 6x + 9$, $g(x) = x + 3$

  C. $f(x) = |x|$, $g(x) = x - 10$

  D. $f(x) = 1/x^2$, $g(x) = 2x - 5$

7A. $(f \circ g \circ h)(x) = 30x$

  B. $(f \circ g \circ h)(x) = x - 1$

  E. $f(x) = x^3$, $g(x) = 1 - x$

  F. $f(x) = \sqrt[3]{x}$, $g(x) = x^2 - x - 8$

  G. $f(x) = |x|^5$, $g(x) = 3x$

  H. $f(x) = x/(4x + 13)$, $g(x) = x + 2$

  C. $(f \circ g \circ h)(x) = x - 14$

  D. $(f \circ g \circ h)(x) = |x|$

Set 91

1A. 1   B. -4 C. -10 D. 0   E. -2 F. 0

  G. 3   H. 3   I. $-2/3$ J. 6   K. 0   L. -4

3A. $f(0) = -6$, but $g(6) \neq 0$ B. $g(-3) = 4$, but $f(4) \neq -3$ C. $f(9) = 1$, but $g(1) \neq 9$

  D. $g(-1) = 1$, but $f(1) \neq -1$ **4**A. 5   B. 3   C. 7   D. -2

5A. $f^{-1}(x) = (x + 5)/2$;
  $\mathbb{D}_{f^{-1}} = \mathbf{R}$

  B. $g^{-1}(x) = (x^2 - 8)/3$;
  $\mathbb{D}_{g^{-1}} = [0, \infty)$

  C. $h^{-1}(x) = (\sqrt[5]{x + 11})/2$;
  $\mathbb{D}_{h^{-1}} = \mathbf{R}$

  D. $f^{-1}(x) = (x^3 - 5)/12$;
  $\mathbb{D}_{f^{-1}} = \mathbf{R}$

  E. $g^{-1}(x) = \sqrt{-x - 7}$;
  $\mathbb{D}_{g^{-1}} = (-\infty, -7]$

  F. $h^{-1}(x) = (x - 4)/(9x + 3)$;
  $\mathbb{D}_{h^{-1}} = \{x \in \mathbf{R} \mid x \neq -1/3\}$

  G. $f^{-1}(x) = 1/x$;
  $\mathbb{D}_{f^{-1}} = \{x \in \mathbf{R} \mid x \neq 0\}$

  H. $g^{-1}(x) = 4 - \sqrt{x + 18}$;
  $\mathbb{D}_{g^{-1}} = [-18, \infty)$

  I. $h^{-1}(x) = 10 - x$;
  $\mathbb{D}_{h^{-1}} = \mathbf{R}$

  J. $f^{-1}(x) = (1 - x^2)/6$;
  $\mathbb{D}_{f^{-1}} = [0, \infty)$

  K. $g^{-1}(x) = (\sqrt[3]{x - 10})/3$;
  $\mathbb{D}_{f^{-1}} = \mathbf{R}$

  L. $h^{-1}(x) = (96 - x^5)/32$;
  $\mathbb{D}_{h^{-1}} = \mathbf{R}$

  M. $f^{-1}(x) = -5 - \sqrt{x - 8}$;
  $\mathbb{D}_{f^{-1}} = [8, \infty)$

  N. $g^{-1}(x) = -x$;
  $\mathbb{D}_{g^{-1}} = \mathbf{R}$

  O. $h^{-1}(x) = 11x/(1 - 6x)$;
  $\mathbb{D}_{f^{-1}} = \{x \in \mathbf{R} \mid x \neq 1/6\}$

  P. $f^{-1}(x) = (-2 + \sqrt{2x + 2})/2$; $\mathbb{D}_{f^{-1}} = [-1, \infty)$

Set 92

1A.

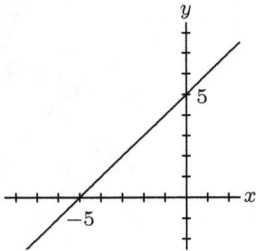

B.

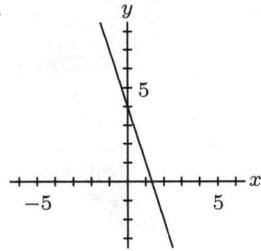

C.

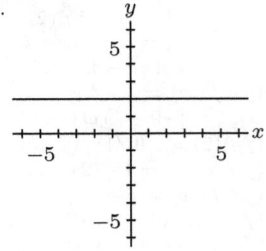

D.

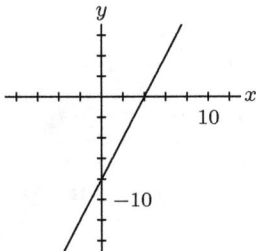

E.

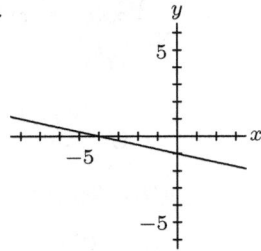

F.

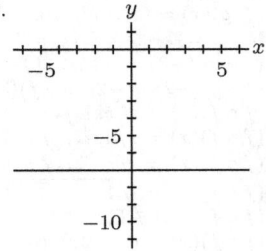

2A. $f(x) = 3x + 14$

  D. $f(x) = (5/2)x - 2$

  B. $f(x) = -9$

  E. $f(x) = (-5/7)x + 5$

  C. $f(x) = -2x + 6$

  F. $f(x) = -x - 8$

5. $-3/4$

Set 93

1A. v: $(7,3)$; a: $x = 7$;
min: 3

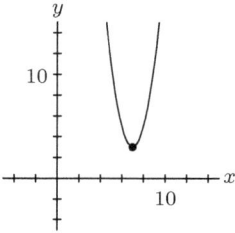

B. v: $(0,-10)$; a: $x = 0$;
max: -10

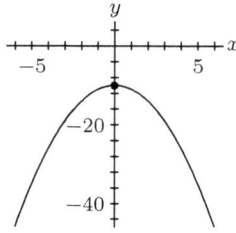

C. v: $(-2,-5)$; a: $x = -2$;
min: -5

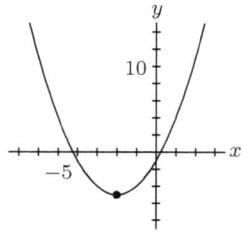

D. v: $(3,-7)$; a: $x = 3$;
min: -7

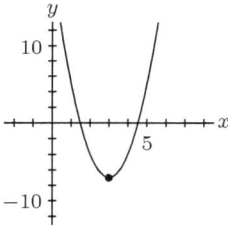

E. v: $(-7, 49/2$; a: $x = -7$;
max: $49/2$

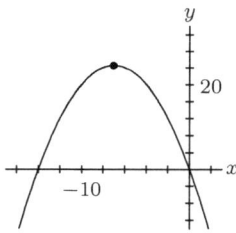

F. v: $(3/8, -105/16$; a: $x = 3/8$;
min: $-105/16$

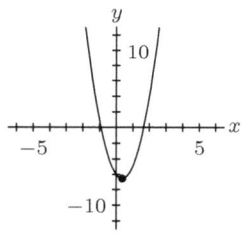

G. v: $(-1,-8)$; a: $x = -1$;
max: -8

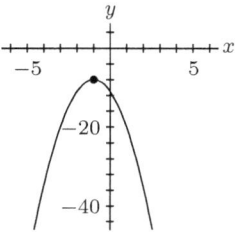

H. v: $(0,0)$; a: $x = 0$;
min: 0

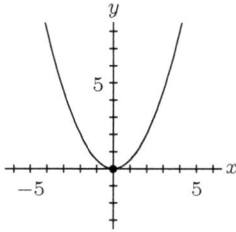

I. v: $(-1/2, -19/2)$; a: $x = -1/2$;
min: $-19/2$

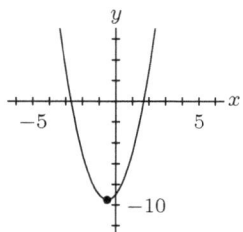

J. v: $(5,-23)$; a: $x = 5$;
min: -23

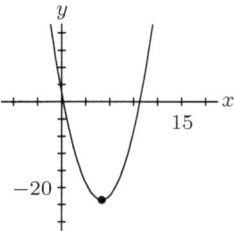

K. v: $(5/2, 29/4)$; a: $x = 5/2$;
max: $29/4$

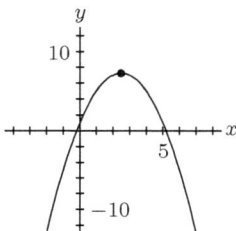

L. v: $(1/3, 17/3)$; a: $x = 1/3$;
min: $17/3$

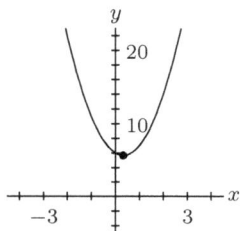

2A. $f(x) = 3(x-2)^2 - 11$ B. $f(x) = -(x+5)^2 + 3$ C. $f(x) = (1/2)(x+1)^2 - 8$
D. $f(x) = -2(x-4)^2 + 1$ E. $f(x) = -(x+4)^2 + 9$

3A. 175 ft at $t = 1.25$ B. 50 yd by 100 yd   C. 15 and 15   D. 5 and 5
E. $27,500 at $x = 50$ F. 65 people   G. 54   H. $(2,1)$

215

Set 94

1A. -3 B. -7 C. 13 D. 1 2A. yes   B. no   C. no   D. yes

3A. $f(x) = 2(x - (-7 + \sqrt{57})/4)(x - (-7 - \sqrt{57})/4)$

B. $g(x) = (x + 2)(x - 3)(x - 5)$ C. $h(x) = (x + 4)^2(3x - 5)$

D. $f(x) = 3(x - 8)(x - (1 + \sqrt{13})/6)(x - (1 - \sqrt{13})/6)$

E. $g(x) = 3(x - (9 + \sqrt{33})/6)(x - (9 - \sqrt{33})/6)$

F. $h(x) = (x + 7)(x + 5)(x + 1)$ G. $f(x) = 2(x + 7)(x - 3)(x - 2)$

H. $g(x) = 2(x + 5)(x - (-2 + \sqrt{10})/2)(x - (-2 - \sqrt{10})/2)$ 5. $-x + 8$ 6. $2/3$

Set 95

1A. $\{\pm 1/4, \pm 1/2, \pm 1, \pm 2\}$ C. $\{\pm 1/3, \pm 2/3, \pm 1, \pm 2, \pm 3, \pm 6\}$

B. $\{\pm 1, \pm 5\}$ D. $\{\pm 1/8, \pm 1/4, \pm 1/2, \pm 1\}$

2A. 3; $(x - 3)(x - (-15 + \sqrt{197})/2)(x - (-15 - \sqrt{197})/2)$

B. $5/2$; $(2x - 5)(x - (1 + \sqrt{13})/2)(x - (1 - \sqrt{13})/2)$ C. -1; $(x + 1)^2(7x + 2)(x - 6)$

D. -2; $(x + 2)(x - (3 + \sqrt{29})/2)(x - (3 - \sqrt{29})/2)$

E. $-8/3$; $(3x + 8)(x - 2 + \sqrt{3})(x - 2 - \sqrt{3})$ F. 2; $(x + 10)(2x - 1)(x - 2)^2$

Set 96

1A. var. in sign in $f(x)$: 3;
var. in sign in $f(-x)$: 0;
# pos. zeros: 3 or 1;
# neg. zeros: 0

B. var. in sign in $f(x)$: 0;
var. in sign in $f(-x)$: 2;
# pos. zeros: 0;
# neg. zeros: 2 or 0

C. var. in sign in $f(x)$: 0;
var. in sign in $f(-x)$: 0;
# pos. zeros: 0;
# neg. zeros: 0

D. var. in sign in $f(x)$: 1;
var. in sign in $f(-x)$: 0;
# pos. zeros: 1;
# neg. zeros: 0

E. var. in sign in $f(x)$: 2;
var. in sign in $f(-x)$: 2;
# pos. zeros: 2 or 0;
# neg. zeros: 2 or 0

F. var. in sign in $f(x)$: 1;
var. in sign in $f(-x)$: 4;
# pos. zeros: 1;
# neg. zeros: 4, 2, or 0

G. var. in sign in $f(x)$: 0;
var. in sign in $f(-x)$: 1;
# pos. zeros: 0;
# neg. zeros: 1

H. var. in sign in $f(x)$: 2;
var. in sign in $f(-x)$: 0;
# pos. zeros: 2 or 0;
# neg. zeros: 0

I. var. in sign in $f(x)$: 6;
var. in sign in $f(-x)$: 1;
# pos. zeros: 6, 4, 2, or 0;
# neg. zeros: 1

Set 97

1A. dom. term: $-x^3$;
$f(x) \to -\infty$ as $x \to \infty$;
$f(x) \to \infty$ as $x \to -\infty$

B. dom. term: $7x^5$;
$f(x) \to \infty$ as $x \to \infty$;
$f(x) \to -\infty$ as $x \to -\infty$

C. dom. term: $-4x^3$;
$f(x) \to -\infty$ as $x \to \infty$;
$f(x) \to \infty$ as $x \to -\infty$

D. dom. term: $-x^4$;
$f(x) \to -\infty$ as $x \to \infty$;
$f(x) \to -\infty$ as $x \to -\infty$

E. dom. term: $56x^4$;
$f(x) \to \infty$ as $x \to \infty$;
$f(x) \to \infty$ as $x \to -\infty$

F. dom. term: $2x^4$;
$f(x) \to \infty$ as $x \to \infty$;
$f(x) \to \infty$ as $x \to -\infty$

G. dom. term: $-25x^9$;
$f(x) \to -\infty$ as $x \to \infty$;
$f(x) \to \infty$ as $x \to -\infty$

H. dom. term: $5x^3$;
$f(x) \to \infty$ as $x \to \infty$;
$f(x) \to -\infty$ as $x \to -\infty$

I. dom. term: $3x^3$;
$f(x) \to \infty$ as $x \to \infty$;
$f(x) \to -\infty$ as $x \to -\infty$

J. dom. term: $-18x^4$; $f(x) \to -\infty$ as $x \to \infty$; $f(x) \to -\infty$ as $x \to -\infty$

Set 98

1A. x-ints: $-2, 1, 3$; y-int: 6;

expression	$-$	$+$	$-$	$+$
$x - 3$	$-$	$-$	$-$	$+$
$x - 1$	$-$	$-$	$+$	$+$
$x + 2$	$-$	$+$	$+$	$+$
	-2	1	3	

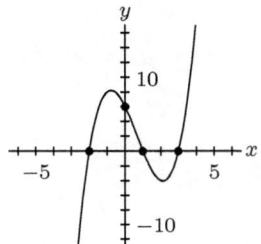

$f(x) \to \infty$ as $x \to \infty$;
$f(x) \to -\infty$ as $x \to -\infty$

B. x-ints: -4, 0, 7; y-int: 0;

expression	$-$	$+$	$+$	$-$
$7-x$	$+$	$+$	$+$	$-$
$x+4$	$-$	$+$	$+$	$+$
$(1/2)x^4$	$+$	$+$	$+$	$+$

$$-4 \qquad 0 \qquad 7$$

$f(x) \to -\infty$ as $x \to \infty$;
$f(x) \to -\infty$ as $x \to -\infty$

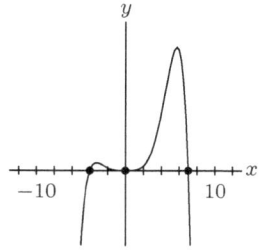

C. x-ints: 0, 3; y-int: 0;

expression	$-$	$+$	$-$
$x-3$	$-$	$-$	$+$
$-7x$	$+$	$-$	$-$

$$0 \qquad 3$$

$f(x) \to -\infty$ as $x \to \infty$;
$f(x) \to -\infty$ as $x \to -\infty$

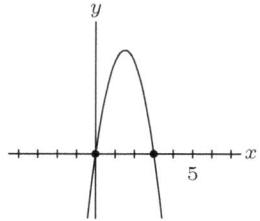

D. x-ints: -3, 2, 3; y-int: 18;

expression	$-$	$+$	$-$	$+$
$x-2$	$-$	$-$	$+$	$+$
$x-3$	$-$	$-$	$-$	$+$
$x+3$	$-$	$+$	$+$	$+$

$$-3 \qquad 2 \qquad 3$$

$f(x) \to \infty$ as $x \to \infty$;
$f(x) \to -\infty$ as $x \to -\infty$

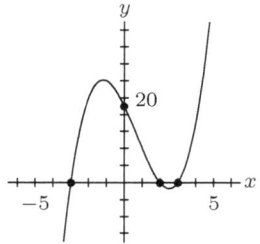

E. x-ints: $-\sqrt{7}$, 0, $\sqrt{7}$; y-int: 0;

expression	$+$	$-$	$+$	$-$
$x-\sqrt{7}$	$-$	$-$	$-$	$+$
$x+\sqrt{7}$	$-$	$+$	$+$	$+$
$-2x$	$+$	$+$	$-$	$-$

$$-\sqrt{7} \qquad 0 \qquad \sqrt{7}$$

$f(x) \to -\infty$ as $x \to \infty$;
$f(x) \to \infty$ as $x \to -\infty$

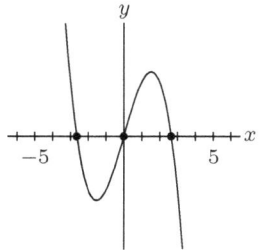

F. x-ints: $-3/5$, 2, 8; y-int: 48;

expression	$-$	$+$	$-$	$+$
$x-8$	$-$	$-$	$-$	$+$
$x-2$	$-$	$-$	$+$	$+$
$5x+3$	$-$	$+$	$+$	$+$

$$-\tfrac{3}{5} \qquad 2 \qquad 8$$

$f(x) \to \infty$ as $x \to \infty$;
$f(x) \to -\infty$ as $x \to -\infty$

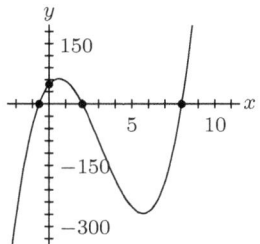

G. x-ints: -5, 0, 2, 4; y-int: 0;

expression		$-$	$+$	$-$	$+$	$-$
$-3x$		$+$	$+$	$-$	$-$	$-$
$x + 5$		$-$	$+$	$+$	$+$	$+$
$x - 2$		$-$	$-$	$-$	$+$	$+$
$x - 4$		$-$	$-$	$-$	$-$	$+$
		-5	0	2	4	

$f(x) \to -\infty$ as $x \to \infty$;
$f(x) \to -\infty$ as $x \to -\infty$

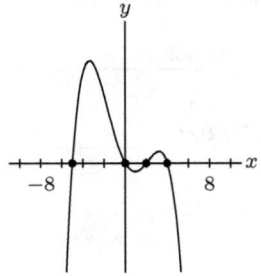

H. x-ints: -1, 3, 8; y-int: -72;

expression	$+$	$-$	$-$	$+$
$(x-3)^2$	$+$	$+$	$+$	$+$
$x + 1$	$-$	$+$	$+$	$+$
$x - 8$	$-$	$-$	$-$	$+$
	-1	3	8	

$f(x) \to \infty$ as $x \to \infty$;
$f(x) \to \infty$ as $x \to -\infty$

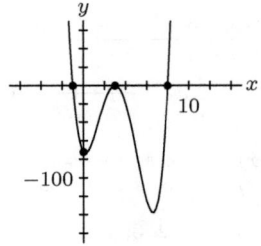

I. x-ints: $5/2$, 10; y-int: $25/4$;

expression	$+$	$-$	$+$
$1/8$	$+$	$+$	$+$
$5 - 2x$	$+$	$-$	$-$
$10 - x$	$+$	$+$	$-$
	$\frac{5}{2}$	10	

$f(x) \to \infty$ as $x \to \infty$;
$f(x) \to \infty$ as $x \to -\infty$

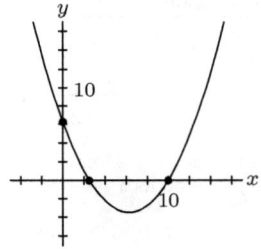

J. x-ints: 0, 5; y-int: 0;

expression	$+$	$-$	$-$
$-x$	$+$	$-$	$-$
$(x-5)^2$	$+$	$+$	$+$
	0	5	

$f(x) \to -\infty$ as $x \to \infty$;
$f(x) \to \infty$ as $x \to -\infty$

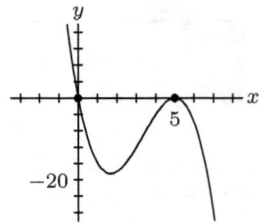

K. x-ints: -4, -3, 4; y-int: -48;

expression	$-$	$+$	$-$	$+$
$x + 3$	$-$	$-$	$+$	$+$
$x + 4$	$-$	$+$	$+$	$+$
$x - 4$	$-$	$-$	$-$	$+$
	-4	-3	4	

$f(x) \to \infty$ as $x \to \infty$;
$f(x) \to -\infty$ as $x \to -\infty$

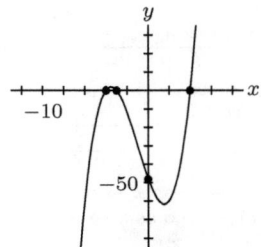

L. x-ints: -3, $7/2$; y-int: -147;

expression	+	−	−
$-(2x-7)^2$	−	−	−
$x+3$	−	+	+

$$-3 \qquad \tfrac{7}{2}$$

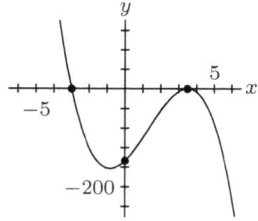

$f(x) \to -\infty$ as $x \to \infty$;
$f(x) \to \infty$ as $x \to -\infty$

Set 99

1A. vert.: $x = -5$; hor.: $y = 0$
  B. vert.: $x = -1/2$, $x = 9$; hor.: $y = 1/2$
  C. vert.: $x = -3$; obl.: $y = -8x + 9$
  D. vert.: $x = -2$; obl.: $y = x + 5$
  E. vert.: $x = -2/3$; hor.: $y = 1/3$
  F. hor.: $y = -2$

  G. vert.: $x = -4$, $x = 4/7$; hor.: $y = 0$
  H. vert.: $x = -1$, $x = -5/6$; hor.: $y = 1/4$
  I. vert.: $x = 5$; obl.: $y = 2x + 5$
  J. vert.: $x = -3/2$, $x = 2$; obl.: $y = -5x + 1$
  K. vert.: $x = -3$; obl.: $y = x - 3$
  L. obl.: $y = x - 2$

Set 100

1A. no x-ints; y-int: -3;

expression	+	−
-12	−	−
$x+4$	−	+

$$-4$$

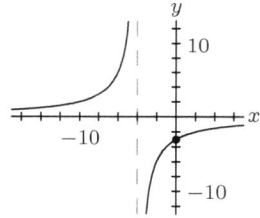

vert. asy.: $x = -4$;
hor. asy.: $y = 0$

B. no x-ints; y-int: 4;

expression	+	+	+
36	+	+	+
$(4x-1)^2$	+	+	+
$(x-3)^2$	+	+	+

$$\tfrac{1}{4} \qquad 3$$

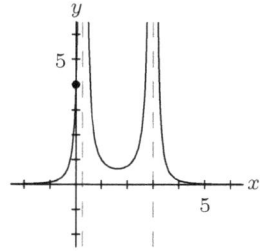

vert. asy.: $x = 1/4$, $x = 3$;
hor. asy.: $y = 0$

C. x-ints: 1, 3; y-int: $-1/3$;

expression	+	−	+	−	+
$x-1$	−	−	+	+	+
$x-3$	−	−	−	+	+
$2x+1$	−	+	+	+	+
$x-9$	−	−	−	−	+

$$-\tfrac{1}{2} \qquad 1 \qquad 3 \qquad 9$$

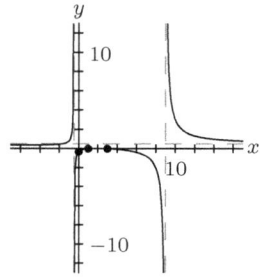

vert. asy.: $x = -1/2$, $x = 9$;
hor. asy.: $y = 1/2$

D. x-ints: $-11/4$, 1; y-int: $22/3$;

expression	+	−	+	−
-2	−	−	−	−
$4x + 11$	−	−	+	+
$x - 1$	−	−	−	+
$x + 3$	−	+	+	+
	-3	$-\frac{11}{4}$	1	

vert. asy.: $x = -3$;
obl. asy.: $y = -8x + 10$

E. x-ints: -6, -1; y-int: 3;

expression	−	+	−	+
$x + 1$	−	−	−	+
$x + 6$	−	+	+	+
$x + 2$	−	−	+	+
	-6	-2	-1	

vert. asy.: $x = -2$;
obl. asy.: $y = x + 5$

F. x-int: -1; y-int: $1/2$;

expression	+	−	+
$x + 1$	−	+	+
$3x + 2$	−	−	+
	-1	$-\frac{2}{3}$	

vert. asy.: $x = -2/3$;
hor. asy.: $y = 1/3$

G. no x-ints; y-int: -2;

expression	−	−
	$\frac{4}{3}$	

hor. asy.: $y = -2$

H. no x-ints; y-int: -1;

expression	+	−	+
40	+	+	+
$2x + 10$	−	+	+
$7x - 4$	−	−	+
	-5	$\frac{4}{7}$	

vert. asy.: $x = -5$, $x = 4/7$;
hor. asy.: $y = 0$

I. no x-ints; y-int: $-2/3$;

expression	+	+	−
-8	−	−	−
$(x+2)^2$	+	+	+
$7x+3$	−	−	+

$$-2 \qquad -\tfrac{3}{7}$$

vert. asy.: $x = -2$, $x = -3/7$;
hor. asy.: $y = 0$

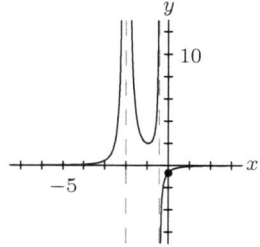

J. x-ints: $-14/3$, 0; y-int: 0;

expression	+	−	+	−	+
x	−	−	−	−	+
$3x+14$	−	+	+	+	+
$6x+5$	−	−	−	+	+
$2x+6$	−	−	+	+	+

$$-\tfrac{14}{3} \qquad -3 \qquad -\tfrac{5}{6} \qquad 0$$

vert. asy.: $x = -3$, $x = -5/6$;
hor. asy.: $y = 1/4$

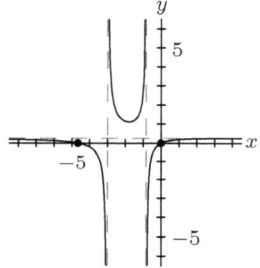

K. x-ints: -2, $9/2$; y-int: $18/5$;

expression	−	+	−	+
$2x-9$	−	−	+	+
$x+2$	−	+	+	+
$x-5$	−	−	−	+

$$-2 \qquad \tfrac{9}{2} \qquad 5$$

vert. asy.: $x = 5$;
obl. asy.: $y = 2x + 5$

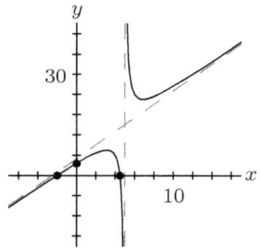

L. x-ints: 1, $6/5$; y-int: 3;

expression	+	−	+	−
$1-x$	+	−	−	−
$5x-6$	−	−	+	+
$x-2$	−	−	−	+

$$1 \qquad \tfrac{6}{5} \qquad 2$$

vert. asy.: $x = 2$;
obl. asy.: $y = -5x + 1$

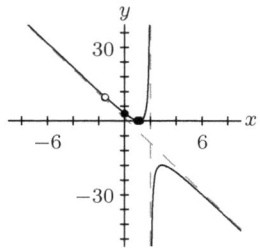

M. no x-ints; y-int: $5/3$;

expression	−	+
x^2+5	+	+
$x+3$	−	+

$$-3$$

vert. asy.: $x = -3$;
obl. asy.: $y = x - 3$

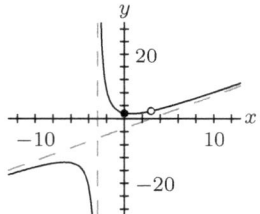

N. x-ints: 2; y-int: -2;

expression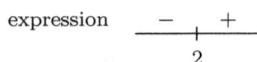

obl. asy.: $y = x - 2$

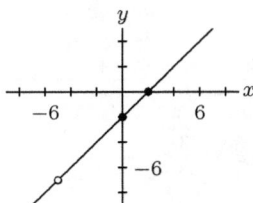

2. 375,000

Set 101

1A. 22.27396　　B. 673.63885　　C. 77.88023　　D. 77.57051　　E. 41.79229　　F. 392.55460

Set 102

1A. $\{2\}$　　　　B. $\{3/10\}$　　C. $\{-11/4\}$　　D. $\{13/2\}$　　E. $\{-1\}$　　F. $\{-2, 9\}$
G. $\{1\}$　　　　H. $\emptyset$　　　　I. $\{3, 7\}$　　　J. $\{-5\}$　　K. $\{1\}$　　L. $\{1/6\}$
M. $\{13/8\}$　　N. $\{-9/4\}$　　O. $\{-5, 3\}$　　P. $\{10\}$　　Q. $\emptyset$　　R. $\{0, 4\}$
2A. 5　　　　　　B. 2　　　　　C. 1/10　　　　D. 9　　　　E. 11　　F. 1/2
3A. $\mathbb{D}$: $\mathbf{R}$; $\mathbb{R}$: $(0, \infty)$;　　　　B. $\mathbb{D}$: $\mathbf{R}$; $\mathbb{R}$: $(0, \infty)$;　　　　C. $\mathbb{D}$: $\mathbf{R}$; $\mathbb{R}$: $(1, \infty)$;
　　asy.: $y = 0$　　　　　　　　asy.: $y = 0$　　　　　　　　asy.: $y = 1$

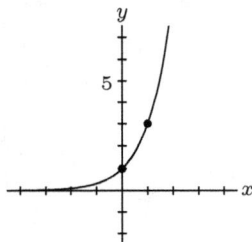　　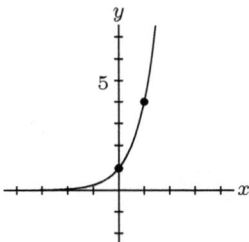　　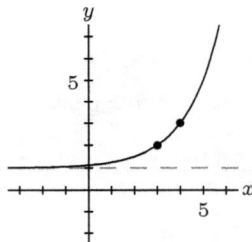

D. $\mathbb{D}$: $\mathbf{R}$; $\mathbb{R}$: $(-\infty, -2)$;　　E. $\mathbb{D}$: $\mathbf{R}$; $\mathbb{R}$: $(0, \infty)$;　　F. $\mathbb{D}$: $\mathbf{R}$; $\mathbb{R}$: $(0, \infty)$;
　　asy.: $y = -2$　　　　　　　asy.: $y = 0$　　　　　　　asy.: $y = 0$

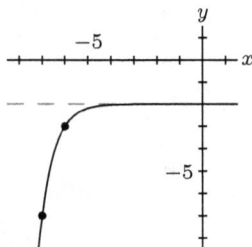　　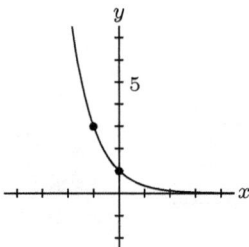　　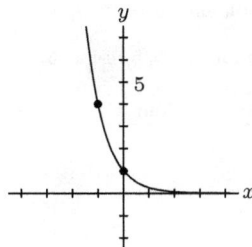

G. $\mathbb{D}$: $\mathbf{R}$; $\mathbb{R}$: $(-\infty, 11)$;　　H. $\mathbb{D}$: $\mathbf{R}$; $\mathbb{R}$: $(-5, \infty)$;
　　asy.: $y = 11$　　　　　　　asy.: $y = -5$

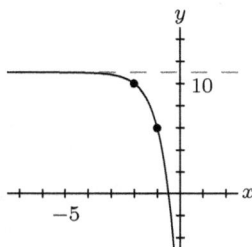　　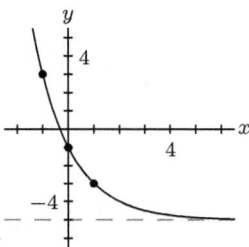

Set 103

1A. 4   B. 1/5   C. -2 D. -3 E. 8   F. 1/2   G. 0   H. 1
I. 6   J. 1/3   K. -3 L. -2 M. 82   N. 7/2   O. 1   P. 0

2A. $2^3 = 8$ B. $e^9 = e^9$ C. $16^{1/2} = 4$ D. $(1/7)^{-1} = 7$ E. $19^{80} = x$
F. $5^4 = 625$ G. $10^2 = 100$ H. $9^{-2} = 1/81$ I. $x^{-4} = 11$ J. $243^{-1/5} = 1/3$

3A. $\log_7 343 = 3$ B. $\log 10{,}000 = 4$ C. $\log_{32} 2 = 1/5$ D. $\log_9 1/9 = -1$
E. $\log_x 50 = 8$ F. $\log_3 243 = 5$ G. $\ln e^6 = 6$ H. $\log_{36} 1/6 = -1/2$
I. $\log_{1/4} 256 = -4$ J. $\log_5 212 = x$ 4A. $(-7/4, \infty)$ B. $(-\infty, -3) \cup (8, \infty)$
C. $\mathbf{R}$ D. $(-\infty, 9/2)$ E. $\{x \in \mathbf{R} \mid x \ne 0\}$ F. $(-6, -5) \cup (0, \infty)$

5A. $\mathbb{D}: (0, \infty)$; $\mathbb{R}: \mathbf{R}$;
asy.: $x = 0$

B. $\mathbb{D}: (-\infty, 0)$; $\mathbb{R}: \mathbf{R}$;
asy.: $x = 0$

C. $\mathbb{D}: (3, \infty)$; $\mathbb{R}: \mathbf{R}$;
asy.: $x = 3$

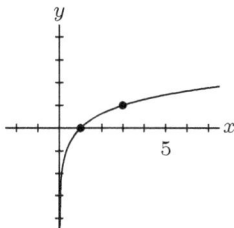 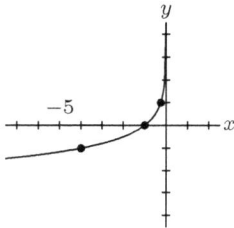 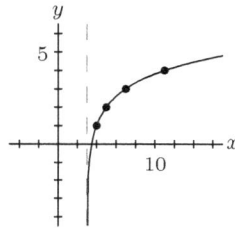

D. $\mathbb{D}: (-6, \infty)$; $\mathbb{R}: \mathbf{R}$;
asy.: $x = -6$

E. $\mathbb{D}: (0, \infty)$; $\mathbb{R}: \mathbf{R}$;
asy.: $x = 0$

F. $\mathbb{D}: (-\infty, 0)$; $\mathbb{R}: \mathbf{R}$;
asy.: $x = 0$

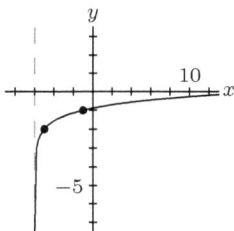 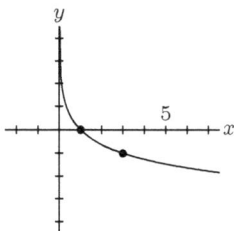 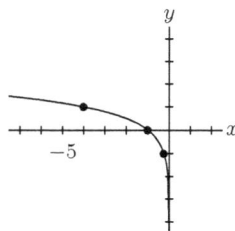

G. $\mathbb{D}: (-2, \infty)$; $\mathbb{R}: \mathbf{R}$;
asy.: $x = -2$

H. $\mathbb{D}: (0, \infty)$; $\mathbb{R}: \mathbf{R}$;
asy.: $x = 0$

6. -3

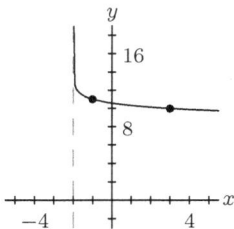 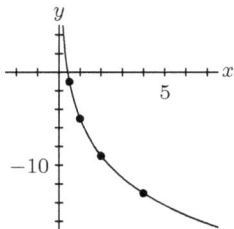

Set 104

1A. $\log_2 13 - \log_2 x$ B. $\log_5 8 + 25 \log_5 x$ C. $3 + 5 \log x + \log y$
D. $4 \log_6 x + \log_6 y - \frac{3}{2} \log_6 z$ E. $\frac{1}{2} + \frac{1}{2} \log_3 (m - 4) + \frac{1}{4} \log_3 (n + 3) + \frac{5}{8} \log_3 p$
F. $\log_9 4 + \log_9 x$ G. $2m + \log_7 n$ H. $2 \log_2 a + 7 \log_2 b - 5$
I. $-3 \ln x - 5 \ln y$ J. $\frac{1}{2} \log_5 x + 3 \log_5 (3y^2 + 10) - \frac{3}{4}$

2A. $\log_7 xy$ B. $\log_{11} x^5/y^2$ C. 2   D. $\ln(x + 5)(3x - 7)$
E. $\log_2 (x + 4)$ F. $\log_8 w/11$ G. $\log 81 m^8 /n^3$ H. 3
I. $\log_4 23x(9x + 1)$ J. $\log_5 (x - 2)$ 3A. $b + c$ B. $3b - 2a$
C. $(a + 2c)/(a + b)$ D. $a - c$ E. $2a + b + c$ F. $(2b - a)/(3a)$

4A. $(\ln 71)/(\ln 9)$;
$(\log 71)/(\log 9)$;
$(\log_2 71)/(\log_2 9)$

B. $(\ln 100)/(\ln 7)$;
$2/(\log 7)$;
$(\log_2 100)/(\log_2 7)$

C. $(\ln 55)/(\ln 4)$;
$(\log 55)/(\log 4)$;
$(\log_2 55)/2$

D. $1/(\ln k)$;
$(\log e)/(\log k)$;
$(\log_2 e)/(\log_2 k)$

E. $(\ln 43)/(\ln 5)$;
$(\log 43)/(\log 5)$;
$(\log_2 43)/(\log_2 5)$

F. $(\ln 63)/(\ln 10)$;
$\log 63$;
$(\log_2 63)/(\log_2 10)$

G. $(\ln 128)/(\ln 11)$;
$(\log 128)/(\log 11)$;
$7/(\log_2 11)$

H. $(\ln b)/(\ln a)$;
$(\log b)/(\log a)$;
$(\log_2 b)/(\log_2 a)$

5A. $(\ln 111)/(\ln 8) \approx 2.26481$

B. $(\ln 3)/(\ln 486) \approx 0.17759$

C. $(\ln 0.00365)/(\ln 11) \approx -2.34081$

D. $(\ln 0.714)/(\ln 10) \approx -0.14630$

E. $(\ln 5)/(\ln 21) \approx 0.52863$

F. $(\ln 250)/(\ln 3) \approx 5.02585$

G. $(\ln 0.0000292)/(\ln 6) \approx -5.82742$

H. $(\ln 969)/(\ln 10) \approx 2.98632$

6A. -1 B. $1/5$ C. -5 D. 2 E. -1 F. -2 G. 3 H. 8

7. $x = b + 3c - 2a$

Set 105

1A. $\{(\ln 77)/(\ln 2)\}$

B. $\{\log 589 - \log 3\}$

C. $\emptyset$

D. $\{\ln 3 - 5\ln 11)/(4\ln 11)\}$

E. $\{-\sqrt{(\ln 91)/(\ln 3)}, \sqrt{(\ln 91)/(\ln 3)}\}$

F. $\{(\ln 7)/(\ln 6)\}$

G. $\{(\ln 347 + 4\ln 5)/(11\ln 5)\}$

H. $\{-4, 7\}$

I. $\{-(\ln 2)/(\ln 7), (\ln 9)/(\ln 7)\}$

J. $\{\log 3\}$

K. $\{(8\ln 13 + 2\ln 6)/(3\ln 6 - 5\ln 13)\}$

L. $\{(-1 + 9\ln 10)/(6 - 2\ln 10)\}$

M. $\{(3 + \log 2)/(\log 5)\}$

N. $\{\ln 17 - \ln 99\}$

O. $\emptyset$

P. $\{(2 + 16\log 7 - \log 23)/(3\log 7)\}$

Q. $\{-(\ln 85)/(\ln 4), (\ln 85)/(\ln 4)\}$

R. $\{(\ln 38 + 9\ln 2)/(5\ln 2)\}$

S. $\{(\ln 2 - \ln 3)/(4\ln 7)\}$

T. $\{-5, -2/3\}$

U. $\{(\ln 12)/(\ln 5)\}$

V. $\{\ln 2, \ln 3\}$

W. $\{(8\ln 2 - 3\ln 3)/(11\ln 2 - 10\ln 3)\}$

X. $\{(7 - 11\ln 10)/(6 - \ln 10)\}$

2A. $(-1, 4)$ B. $(7, -2)$

Set 106

1A. $\{125\}$ B. $\{19\}$ C. $\{3\}$ D. $\{-7\}$ E. $\{-47/5\}$ F. $\{6\}$

G. $\{6\}$ H. $\{8\}$ I. $\{3\}$ J. $\{1, 100\}$ K. $\{10\}$ L. $\{81\}$

M. $\{-5, 5\}$ N. $\{121/5\}$ O. $\{3/2\}$ P. $\{57/5\}$ Q. $\{5\}$ R. $\{4\}$

S. $\{7\}$ T. $\{5/2, 3\}$ U. $\{1, e^{-3\sqrt{3}}, e^{3\sqrt{3}}\}$ V. $\{6\}$

2A. $(\sqrt[3]{4}, 4\sqrt[3]{4})$ B. $(\sqrt[5]{7}, (\sqrt[5]{7})^4)$

Set 107

1A. $((\ln 39)/(\ln 2), \infty)$

B. $\emptyset$

C. $[(-5 - \log 11 + 5\log 7)/(2 - 2\log 7), \infty)$

D. $(-\infty, -(\ln 8 + 3\ln 6)/(\ln 6))$

E. $[(4\ln 5)/(3\ln 5 - \ln 21), \infty)$

F. $[-2, 6]$

G. $(-\infty, (\ln 4 + \ln 5)/(8\ln 5))$

H. $\mathbf{R}$

I. $[2, \infty)$

J. $(-\infty, (40\ln 3 - 10\ln 4)/(\ln 3))$

K. $(-\infty, (\ln 3)/(2\ln 3 + \ln 17)]$

L. $(-\infty, -6] \cup [-1, \infty)$

Set 108

1A. $(8, 33)$

B. $[159919/20000, 8)$

C. $(-\infty, -4) \cup (10, \infty)$

D. $[-5, -3) \cup (3, 5]$

E. $(6, 9)$

F. $(18, \infty)$

G. $(5, 23/3)$

H. $[-6249/3125, \infty)$

I. $(-11, -9) \cup (-7, -5)$

J. $(-\infty, -7] \cup [7, \infty)$

K. $[2, \infty)$

L. $(8/3, 19]$

Set 109

1A. $\$910.74$ B. $\$909.70$ C. $\$903.06$ D. $\$911.01$ E. $\$895.42$ F. $\$907.01$

G. $\$910.43$ H. $\$911.06$ **2**A. $\$60,091.52$ B. $\$14,913.28$

C. $\$27,126.40$ D. $\$12,213.01$ E. $\$33,102.04$ F. $\$356,328.16$

3A. 2052 B. 2037 C. 17.36 years   D. 9.29 years   E. 10.38% F. 22.48%

G. 8.11% H. 6.72% I. $\$549,085.81$ J. $\$922.40$ K. $\$100$ at 5% L. 6% monthly

4A. 36 years   B. 4 years   C. 8 years   D. 12 years

Set 110

1A. 34,000   B. 0.051   C. 53,805   2A. 981 oz   B. -0.18   C. 78.93 oz
3A. 9.63 mo   B. 2.73%   C. 26.35 mo   4A. 2.35 days   B. 490.29   C. 7.81 days
5. 556.04 min   6. 33.71 days   7A. $Q_t = 250e^{0.011363t}$ B. 324.64 days
C. 1,000,991   8A. $Q_t = 80e^{-0.01756t}$ B. 199.69 hours   C. 22.59 mg

Set 111

1A. $p = 4;\ a = 3$ B. $p = 4;\ a = 3/2$ C. $p = 5;\ a = 7/2$ D. $p = 10;\ a = 5/2$
2A. 5   B. $-\pi$ C. -8 D. e
3A. $p_g = 2;\ a_g = 21$ B. $p_g = 2;\ a_g = 8$ C. $p_g = 3;\ a_g = 8$ D. $p_g = 4\pi/7;\ a_g = 50$

Set 112

1A. $(x-3)^2 + (y-8)^2 = 49$ B. $x^2 + (y+7)^2 = 1$ C. $(x+1)^2 + (y+2)^2 = 25$
D. $(x+3)^2 + (y-7)^2 = 29$ E. $(x-2)^2 + (y+3)^2 = 9$ F. $(x+2)^2 + (y-6)^2 = 81$
G. $(x-8)^2 + y^2 = 16$ H. $(x-4)^2 + (y-11)^2 = 25$ I. $(x+4)^2 + (y-9)^2 = 289$
J. $(x+8)^2 + (y+5)^2 = 64$ K. $(x-2)^2 + (y+\frac{5}{2})^2 = \frac{169}{4}$ L. $(x-3)^2 + (y+5)^2 = 5$
2A. c: $(8,-1)$; r: 3   B. c: $(3,4)$; r: 9   C. c: $(0,0)$; r: 6

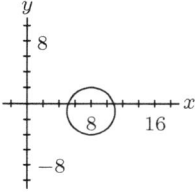
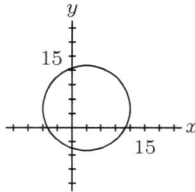
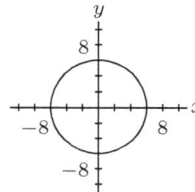

D. c: $(0,-7)$; r: $\sqrt{10}$ E. c: $(-2,5)$; r: 4   F. c: $(-9,-6)$; r: 7

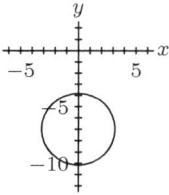
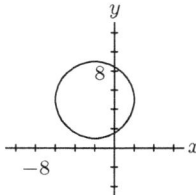
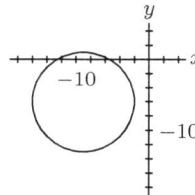

G. c: $(5,0)$; r: $\sqrt{17}$ H. c: $(4,10)$; r: 2   5A.

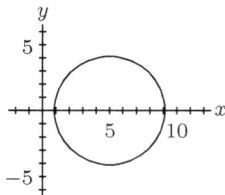
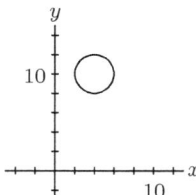
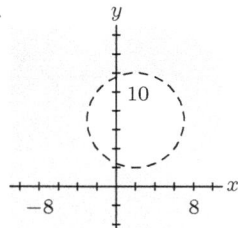

B.   C.   D.

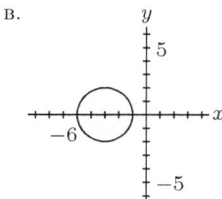
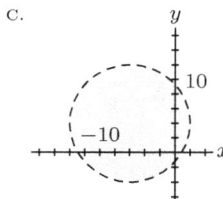
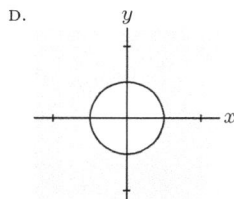

3A. $(x+6)^2 + (y-5)^2 = 81$;
  c: $(-6,5)$; r: 9
B. $x^2 + (y+10)^2 = 64$;
  c: $(0,-10)$; r: 8
C. $(x-2)^2 + (y+5/2)^2 = 15$;
  c: $(2,-5/2)$; r: $\sqrt{15}$
D. $(x-5/3)^2 + (y-2/3)^2 = 50/9$;
  c: $(5/3,2/3)$; r: $5\sqrt{2}/3$

E. $(x-8)^2 + (y-1)^2 = 36$;
  c: $(8,1)$; r: 6
F. $(x+7)^2 + y^2 = 121$;
  c: $(-7,0)$; r: 11
G. $(x+24/9)^2 + (y+1)^2 = 2$;
  c: $(-24/9,-1)$; r: $\sqrt{2}$
H. $(x+11/4)^2 + (y-3/4)^2 = 49/8$;
  c: $(-11/4,3/4)$; r: $7\sqrt{2}/4$

4A. yes   B. no   C. yes   D. yes   E. yes   F. no   **6**. 25π

Set 113
1A. $\angle E$, $\angle DEF$ B. $\angle P$, $\angle SPR$

Set 114
1A. 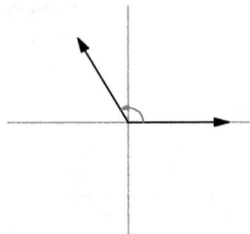   B. 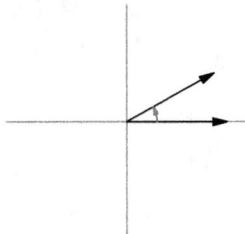   C.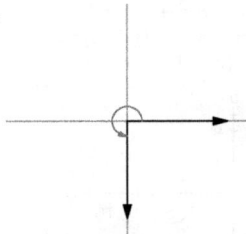

D. 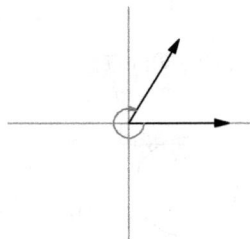   E. 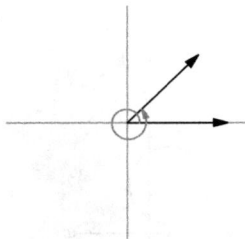   F.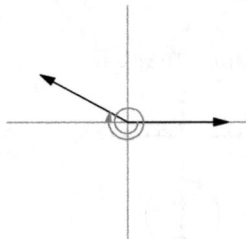

G. 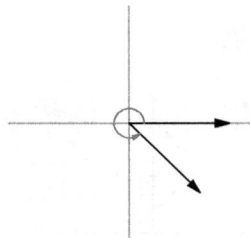   H. 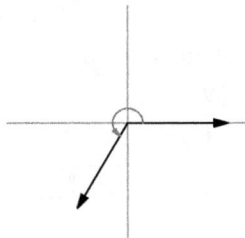   I.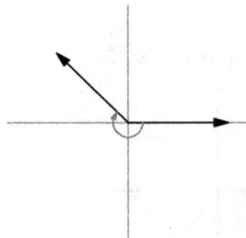

J. 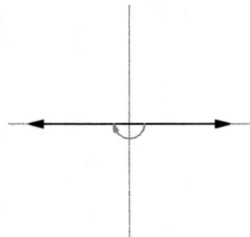   K. 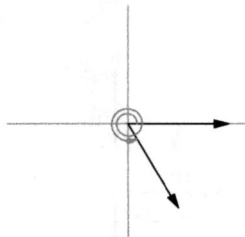   L.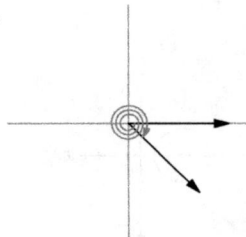

2A. $-510°, -150°, 570°, 930°$ B. $-920°, -200°, 160°, 520°$ C. $-665°, -305°, 55°, 775°$
 D. $-778°, -418°, 302°, 662°$ 3A. $120°$ B. $260°$ C. $280°$ D. $215°$
4A. $55° 14' 30''$ B. $16° 49' 51''$ C. $-38° 24' 8''$ D. $429° 9' 56''$
5A. $43.0436°; 2,582.6167'; 154,957''$ C. $20.7589°; 1,245.5333'; 74,732''$
 B. $100.6656°; 6,039.9333'; 362,396''$ D. $-3.9878°; -239.2667'; -14,356''$

Set 115

1A. 6   B. $15\pi/2$ cm   C. 15 mm   D. 1/7   E. 99π yd   F. 7/15 in
2A.   B.   C.

D.   E.   F.

G.   H.   I.

J.   K.   L.

3A. $-7\pi/3, -\pi/3, 11\pi/3, 17\pi/3$ C. $-29\pi/6, -17\pi/6, 7\pi/6, 19\pi/6$
 B. $-23\pi/4, -7\pi/4, \pi/4, 9\pi/4$ D. $-7\pi/2, -3\pi/2, \pi/2, 5\pi/2$
4A. $7\pi/4$ B. $\pi/2$ C. $5\pi/6$ D. $4\pi/3$ E. π F. $12\pi/7$

Set 116

1A. $\pi/3$ B. $7\pi/6$ C. $-4\pi/3$ D. $-\pi/4$ E. 3π F. $-29\pi/4$
  G. $7\pi/4$ H. $5\pi/6$ I. $-11\pi/6$ J. $-5\pi/4$ K. $-11\pi/2$ L. $9\pi/4$

2A. $330°$ B. $225°$ C. $-315°$ D. $-150°$ E. $-1440°$ F. $600°$
  G. $120°$ H. $15°$ I. $-210°$ J. $-60°$ K. $270°$ 3. 3,423 miles

Set 117

1A. $\pi/3$ B. $\pi/6$ C. $3\pi/11$ D. $4\pi/9$ E. $\pi/4$ F. $\pi/3$ G. n/a   H. $2\pi/5$
  I. $\pi/4$ J. $\pi/7$ K. $\pi/4$ L. n/a   M. $2\pi/7$ N. $2\pi/5$ O. $\pi/6$ P. $\pi/4$
  Q. $\pi/8$ R. $4\pi/11$ S. $\pi/6$ T. $4\pi/9$

Set 118

1A. $-3/5$ B. $-5/4$ C. $4/3$ D. $12/13$ E. $-13/5$ F. $-5/12$
  G. undefined   H. 0   I. 0   J. $3/4$ K. $-5/3$ L. $-4/5$
  M. $-12/5$ N. $13/12$ O. $-5/13$ P. undefined   Q. 1   R. 1

Set 119

1A. $1/2$ B. 1   C. $\sqrt{3}$ D. 1   E. 1   F. $\sqrt{2}/2$ G. $\sqrt{3}$ H. 0
  I. 0   J. $2\sqrt{3}/3$ K. $2\sqrt{3}/3$ L. 1   M. $1/2$ N. $\sqrt{3}/3$
  O. $\sqrt{2}$ P. 1   Q. $\sqrt{3}/3$ R. 1   S. 2   T. $\sqrt{3}/2$
  U. 0   V. $\sqrt{3}/2$ W. $\sqrt{2}$ X. 0   Y. 2   Z. undefined

Set 120

1A. $-\sqrt{2}/2$ B. $-1/2$ C. -1 D. $-2\sqrt{3}/3$ E. -1 F. $-\sqrt{3}/2$
  G. $1/2$ H. $-\sqrt{2}/2$ I. 0   J. 1   K. $-\sqrt{3}$ L. $-\sqrt{2}/2$
  M. undefined   N. $\sqrt{3}/2$ O. -1 P. $-\sqrt{3}/2$ Q. $-\sqrt{2}$ R. $-\sqrt{3}/3$
  S. undefined   T. $\sqrt{2}/2$ U. $\sqrt{3}/2$ V. 0   W. -1 X. $\sqrt{3}$
  Y. $\sqrt{3}/2$ Z. 1

Set 121

1A. $p = 2\pi$; $a = 1$ B. $p = 2\pi$; $a = 5$ C. $p = 3\pi/2$; $a = 1$

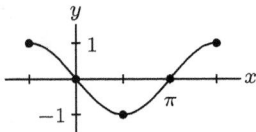   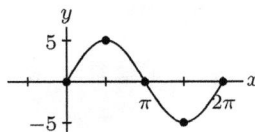   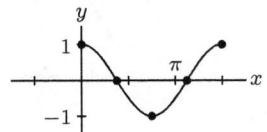

D. $p = \pi$; $a = 1/8$ E. $p = 2$; $a = 2$ F. $p = 2\pi$; $a = 1$

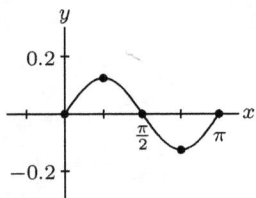   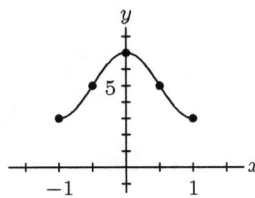   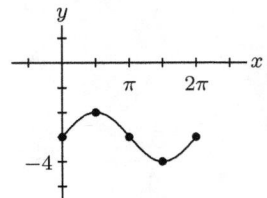

G. $p = 2\pi$; $a = 1/2$ H. $p = \pi/3$; $a = 1$ I. $p = 8\pi$; $a = 3$

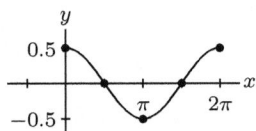   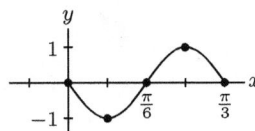   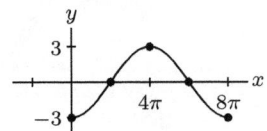

2A. $p = \pi$; $a = 3$;
   $A = 3$, $B = 2$, $C = \pi/4$, $D = 0$
   $f(x) = 3\cos(2(x - \pi/4))$
B. $p = 6\pi$; $a = 2$;
   $A = 2$, $B = 1/3$, $C = -2\pi$, $D = 1$
   $f(x) = 2\cos((x + 2\pi)/3) + 1$

C. $p = 2\pi$; $a = 4$;
   $A = 4$, $B = 1$, $C = \pi$, $D = 0$
   $f(x) = 4\cos(x - \pi)$
D. $p = \pi/2$; $a = 5$;
   $A = 5$, $B = 4$, $C = \pi/8$, $D = -1$
   $f(x) = 5\cos(4(x - \pi/8)) - 1$

3. $A: (3\pi/2, -1)$, $B: (-\pi, 0)$, $C: (4\pi, 0)$, $D: (-7\pi/2, 1)$, $E: (\pi, 0)$

Set 122

1A. $p = \pi$

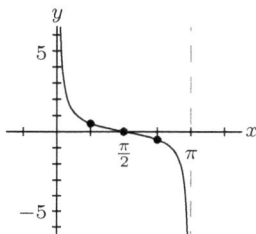

B. $p = 2\pi$

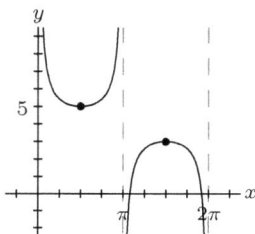

C. $p = \pi$

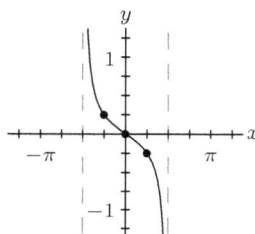

D. $p = 2\pi$

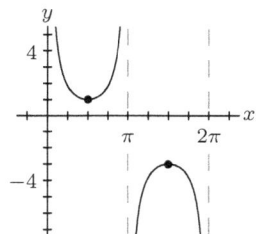

E. $p = \pi/3$

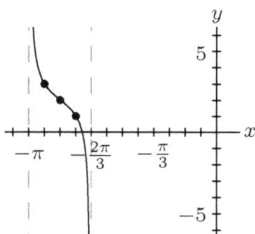

F. $p = \pi$

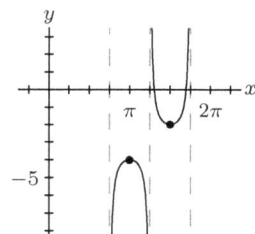

G. $p = 2\pi$

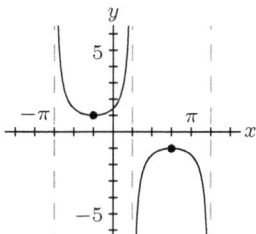

H. $p = \pi/3$

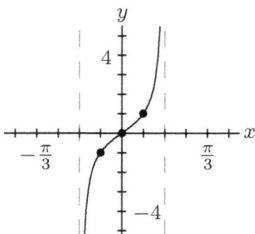

I. $p = \pi$

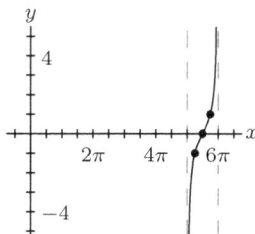

J. $p = \pi$

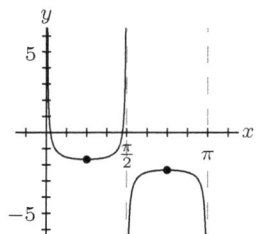

K. $p = 2\pi$

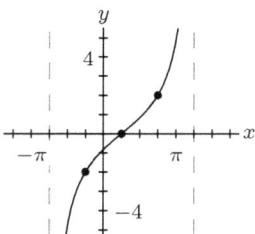

L. $p = 2\pi$

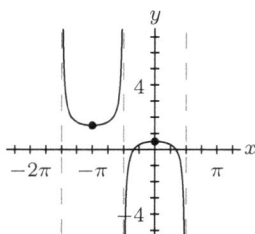

Set 123

1A. $\cos\theta = 4/5$; $\sin\theta = 3/5$; $\tan\theta = 3/4$;
$\sec\theta = 5/4$; $\csc\theta = 5/3$; $\cot\theta = 4/3$
C. $\cos\theta = 3/4$; $\sin\theta = \sqrt{7}/4$; $\tan\theta = \sqrt{7}/3$;
$\sec\theta = 4/3$; $\csc\theta = 4/\sqrt{7}$; $\cot\theta = 3/\sqrt{7}$
B. $\cos\theta = 3/\sqrt{10}$; $\sin\theta = 1/\sqrt{10}$; $\tan\theta = 1/3$;
$\sec\theta = \sqrt{10}/3$; $\csc\theta = \sqrt{10}$; $\cot\theta = 3$
D. $\cos\theta = 15/17$; $\sin\theta = 8/17$; $\tan\theta = 8/15$;
$\sec\theta = 17/15$; $\csc\theta = 17/8$; $\cot\theta = 15/8$

2A. $y = 5\sqrt{3}$; $r = 10$ B. $x = 4\sqrt{2}$; $y = 4\sqrt{2}$ C. $y = 7$; $r = 7\sqrt{2}$ D. $y = 3\sqrt{3}$; $r = 6\sqrt{3}$

3A. 6.9 ft   B. 33.8 yd   C. 50.8 ft   D. 6.2 ft   E. 3.9 ft   F. 482.2 ft

4A. $\cos\theta = 3/5$; $\sin\theta = 4/5$; $\tan\theta = 4/3$;
$\sec\theta = 5/3$; $\csc\theta = 5/4$; $\cot\theta = 3/4$
D. $\cos\theta = \sqrt{35}/6$; $\sin\theta = 1/6$; $\tan\theta = 1/\sqrt{35}$;
$\sec\theta = 6/\sqrt{35}$; $\csc\theta = 6$; $\cot\theta = \sqrt{35}$
B. $\cos\theta = 2/3$; $\sin\theta = \sqrt{5}/3$; $\tan\theta = \sqrt{5}/2$;
$\sec\theta = 3/2$; $\csc\theta = 3/\sqrt{5}$; $\cot\theta = 2/\sqrt{5}$
E. $\cos\theta = 1/\sqrt{65}$; $\sin\theta = 8/\sqrt{65}$; $\tan\theta = 8$;
$\sec\theta = \sqrt{65}$; $\csc\theta = \sqrt{65}/8$; $\cot\theta = 1/8$
C. $\cos\theta = 5/13$; $\sin\theta = 12/13$; $\tan\theta = 12/5$;
$\sec\theta = 13/5$; $\csc\theta = 13/12$; $\cot\theta = 5/12$
F. $\cos\theta = 2/9$; $\sin\theta = \sqrt{77}/9$; $\tan\theta = \sqrt{77}/2$;
$\sec\theta = 9/2$; $\csc\theta = 9/\sqrt{77}$; $\cot\theta = 2/\sqrt{77}$

5. angles: $30°$, $60°$; legs: 7, $7\sqrt{3}$, 14; side: 14 **7.** $30\sqrt{3}$

Set 124

2A. $\csc\theta$ B. 1   C. 1   D. 0   E. 1   F. 0

3A. $-3/5$ B. $8/15$ C. $-\sqrt{21}/5$ D. $12/5$ E. $-\sqrt{17}$ F. $\sqrt{5}/2$

Set 125

2A. $1/2$ B. $\sqrt{2}/2$ C. 0   D. $\cos 5\theta$ E. 0   F. 0

3A. $(\sqrt{6}-\sqrt{2})/4$ B. $(\sqrt{6}-\sqrt{2})/4$ C. $(\sqrt{2}-\sqrt{6})/4$ D. $(\sqrt{6}+\sqrt{2})/4$

4. $\cos(\alpha-\beta) = 36/85$;
$\sin(\alpha+\beta) = 13/85$
5. $\cos(\alpha+\beta) = (24 - 5\sqrt{5})/39$;
$\sin(\alpha-\beta) = (-10 + 12\sqrt{5})/39$

7A. $(7\sqrt{2})\sin(x+\pi/4)$ B. $6\sin(x + 5\pi/6)$ C. $8\sin(x + 4\pi/3)$ D. $(2\sqrt{2})\sin(x + 7\pi/4)$

Set 126

2A. $527/625$ B. $336/625$ C. $336/527$ D. $7\sqrt{2}/10$ E. $\sqrt{2}/10$ F. $1/7$

3A. $\dfrac{\sqrt{2+\sqrt{2}}}{2}$ B. $2 - \sqrt{3}$ C. $\dfrac{\sqrt{2-\sqrt{3}}}{2}$ D. $-\dfrac{\sqrt{2-\sqrt{3}}}{2}$ E. $\dfrac{\sqrt{2-\sqrt{3}}}{2}$ F. $1 - \sqrt{2}$

4. $\cos(\alpha/2) = \sqrt{17}/17$;
$\sin(\alpha/2) = 4\sqrt{17}/17$;
$\tan(\alpha/2) = 4$
5. $\cos(2\beta) = -119/169$;
$\sin(2\beta) = -120/169$;
$\tan(2\beta) = 120/119$
6. $\sqrt{1+c}$
7. $-15/8$

8A. $\dfrac{\cos 4\theta - 4\cos 2\theta + 3}{8}$ B. $\dfrac{(\sin\theta)(1 - \cos 4\theta)}{8}$ C. $\dfrac{(\cos\theta)(\cos 4\theta + 4\cos 2\theta + 3)}{8}$

D. $(5 - 7\cos 2\theta + 3\cos 4\theta - \cos 4\theta \cos 2\theta)/(5 + 7\cos 2\theta + 3\cos 4\theta + \cos 4\theta \cos 2\theta)$

9. angles: $22.5°$, $67.5°$; legs of triangle: $3\sqrt{2-\sqrt{2}}$, $3\sqrt{2+\sqrt{2}}$, 6; side of oct: $6\sqrt{2-\sqrt{2}}$

10A. $3/5$ B. $4/5$

Set 127

2A. $(2 - \sqrt{2})/4$ B. $(-2 + \sqrt{3})/4$ C. $(\cos 3\alpha - \cos 7\alpha)/2$ D. $(\cos 2\theta)/2$
E. $(-2 + \sqrt{3})/4$ F. $-1/4$ G. $(\sin 6t - \sin 2t)/2$ H. $(2\sin 2\theta - \sqrt{3})/4$

3A. $-\sqrt{2}/2$ B. 0   C. $-2\cos\theta$ D. $-\sqrt{2}/2$
E. $-\left(\sqrt{4 - 2\sqrt{2}}\right)/2$ F. $2\cos(\theta - 5\pi/4)$

Set 128

1A. $\pi/6$ B. $-\pi/6$ C. $3\pi/4$ D. $-\pi/2$ E. 0   F. $\pi/6$
G. π H. $\pi/6$ I. 0.2 J. π K. $3\pi/4$ L. undefined
M. $(\sqrt{6}+\sqrt{2})/4$ N. $\pi/4$ O. $\pi/2$ P. $-\pi/3$ Q. $\pi/3$
R. $-\pi/6$ S. $\pi/4$ T. $-\pi/4$ U. $\pi/6$ V. 8   W. $-\pi/8$
X. $-\pi/6$ Y. 0   Z. $\sqrt{3}$ **2**A. $8/17$ B. $7/\sqrt{33}$ C. $\sqrt{29}/5$
D. $2\sqrt{2}/3$ **3**A. $\sqrt{4 - x^2}/2$ B. $\sqrt{x^2 - 49}/7$ C. $\sqrt{1 + x^2}/x$ D. $\sqrt{9 - x^2}/3$

4A.

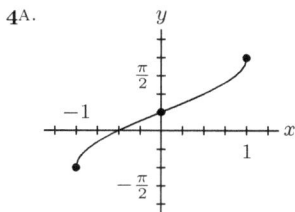

B.

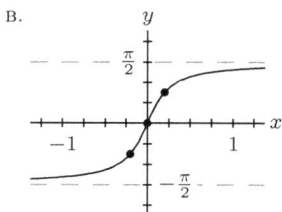

C.

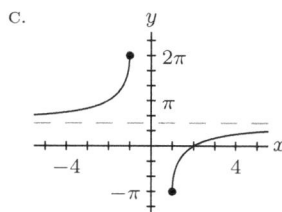

D.

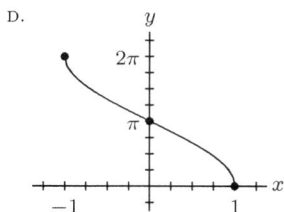

E.

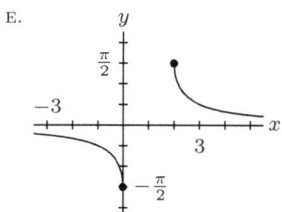

F.

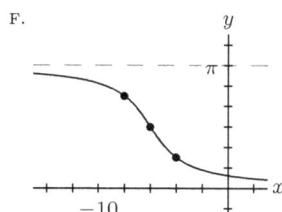

Set 129

1A. $\{\pi/3 + 2\pi k \mid k \in \mathbf{Z}\} \cup \{2\pi/3 + 2\pi k \mid k \in \mathbf{Z}\}$

B. $\{2\pi/3 + 2\pi k \mid k \in \mathbf{Z}\} \cup \{4\pi/3 + 2\pi k \mid k \in \mathbf{Z}\}$

C. $\{7\pi/6 + 2\pi k \mid k \in \mathbf{Z}\} \cup \{11\pi/6 + 2\pi k \mid k \in \mathbf{Z}\} \cup \{3\pi/2 + 2\pi k \mid k \in \mathbf{Z}\}$

D. $\{\pi/3 + 2\pi k \mid k \in \mathbf{Z}\} \cup \{5\pi/3 + 2\pi k \mid k \in \mathbf{Z}\} \cup \{\pi + 2\pi k \mid k \in \mathbf{Z}\}$

E. $\{3\pi/4 + \pi k \mid k \in \mathbf{Z}\}$ F. $\{\pi/4 + \pi k/2 \mid k \in \mathbf{Z}\}$ G. $\{2\pi k \mid k \in \mathbf{Z}\}$

H. $\{\pi/4 + 2\pi k \mid k \in \mathbf{Z}\} \cup \{7\pi/4 + 2\pi k \mid k \in \mathbf{Z}\}$ I. $\{\pi k \mid k \in \mathbf{Z}\}$

J. $\{7\pi/6 + 2\pi k \mid k \in \mathbf{Z}\} \cup \{11\pi/6 + 2\pi k \mid k \in \mathbf{Z}\}$

K. $\{\pi/2 + \pi k \mid k \in \mathbf{Z}\} \cup \{\pi/4 + \pi k \mid k \in \mathbf{Z}\}$ L. $\{\pi/3 + \pi k \mid k \in \mathbf{Z}\} \cup \{2\pi/3 + \pi k \mid k \in \mathbf{Z}\}$

M. $\{\pi/2 + \pi k \mid k \in \mathbf{Z}\}$ N. $\{\pi/6 + \pi k \mid k \in \mathbf{Z}\} \cup \{5\pi/6 + \pi k \mid k \in \mathbf{Z}\}$

2A. $\{75° + 180°k \mid k \in \mathbf{Z}\} \cup \{105° + 180°k \mid k \in \mathbf{Z}\}$; $\{75°, 105°, 255°, 285°\}$

B. $\{240° + 360°k \mid k \in \mathbf{Z}\}$; $\{240°\}$

C. $\{52.5° + 90°k \mid k \in \mathbf{Z}\} \cup \{82.5° + 90°k \mid k \in \mathbf{Z}\}$; $\{52.5°, 82.5°, 142.5°, 172.5°, 232.5°,$

D. $\{90° + 1080°k \mid k \in \mathbf{Z}\} \cup \{450° + 1080°k \mid k \in \mathbf{Z}\}$; $\{90°\}$ $262.5°, 322.5°, 352.5°\}$

E. $\{45° + 90°k \mid k \in \mathbf{Z}\}$; $\{45°, 135°, 225°, 315°\}$

F. $\{20° + 60°k \mid k \in \mathbf{Z}\}$; $\{20°, 80°, 140°, 200°, 260°, 320°\}$

3A. $\{3.8103, 5.6144\}$ B. $\{0.2864, 2.8552\}$ C. $\{0.7043, 2.2751, 3.8459, 5.4167\}$

D. $\{1.7518, 4.5314\}$ E. $\{3.0221, 6.1637\}$ F. $\{0.8163, 5.4669\}$ **4**A. $\{24.49°, 335.51°\}$

B. $\{104.60°, 165.40°, 284.60°, 345.40°\}$ C. $\{53.50°, 233.50°\}$ D. $\{201.10°, 338.90°\}$

E. $\{23.72°, 68.72°, 113.72°, 158.72°, 203.72°, 248.72°, 293.72°, 338.72°\}$

F. $\{34.29°, 85.71°, 154.29°, 205.71°, 274.29°, 325.71°\}$

5A. $112.62°$ B. $33.99°$ C. $208.07°$ D. $309.23°$ **6**A. $(\sqrt{34})\sin(x + 2.1112)$

B. $6\sin(x + 5.0943)$ C. $(3\sqrt{13})\sin(x + 0.5880)$ D. $3\sin(x + 4.2215)$

Set 130

1A. 19.66 B. 5.83 C. 10.33 D. 17.54 **2**A. $5\sqrt{2}/4$ B. $3\sqrt{3}/4$ C. $3\sqrt{3}/4$ D. $\dfrac{4\pi - 3\sqrt{3}}{4}$

Set 131

1A. $25\pi/6$ B. $28\pi/3$ **2.** $(\pi - 2)/4$ **3.** 3π in$^2 \approx 9.42$ in^2

Set 132

1A. 7.03 B. 17.17 C. 5.98 D. $24.48°$ E. $18.99°$ F. $63.47°$ or $116.53°$ **2.** 632.2 yds

Set 133

1A. 3.07 B. 21.84 C. 15.23 D. $48.98°$ E. $160.53°$

F. $\alpha = 52.62°$; $\beta = 15.36°$; $\gamma = 112.02°$ **2.** 14 **3.** $(36 + 25\sqrt{3})/4$

Set 134

1A. area = 20.98;
θ = 50.98°

B. area = 32.97;
θ = 109.62°

C. area = 18.33;
θ = 113.58°

D. area = 32.84;
θ = 29.84°

Set 135

1A. circle   B. ellipse   C. parabola   D. hyperbola

Set 136

1A. $y^2 = 16x$ B. $x^2 = -4y$ C. $x^2 = -8y$ D. $y^2 = 12x$ E. $x^2 = 20y$ F. $y^2 = -24x$
G. $y^2 = -28x$ H. $x^2 = -32y$ I. $x^2 = 20y$ J. $y^2 = -8x$ K. $y^2 = 12x$ L. $x^2 = 8y$
M. $y^2 = 8x$ N. $x^2 = -5y$ O. $x^2 = 11y$ P. $y^2 = -10x$

2A. focus: $(0, -3/2)$; directrix: $y = 3/2$; vertex: $(0,0)$; focal diameter: 6
B. focus: $(-35/2, 6)$; directrix: $x = -37/2$; vertex: $(-18, 6)$; focal diameter: 2
C. focus: $(3/2, -4/5)$; directrix: $y = -33/10$; vertex: $(3/2, -41/20)$; focal diameter: 5
D. focus: $(9/4, 0)$; directrix: $x = -9/4$; vertex: $(0,0)$; focal diameter: 9
E. focus: $(-1, 4)$; directrix: $y = 2$; vertex: $(-1, 3)$; focal diameter: 4
F. focus: $(-67/32, -5/2)$; directrix: $x = 61/32$; vertex: $(-3/32, -5/2)$; focal diameter: 8

3A. $x^2 = 16y$ B. $(y-5)^2 = 4(x-3)$ C. $(x-8)^2 = 28(y-5)$

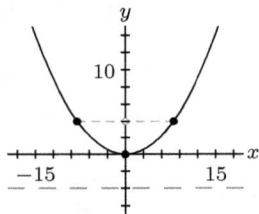 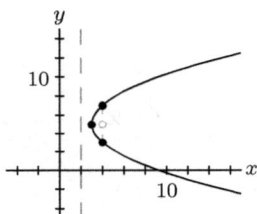 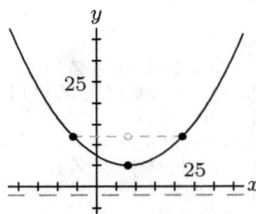

D. $y^2 = -12x$ E. $(x+8)^2 = -12(y-2)$ F. $(y-1)^2 = -12(x+6)$

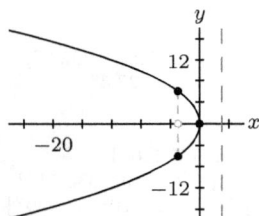 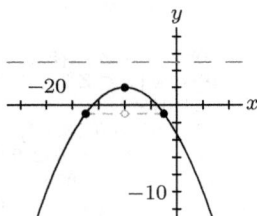 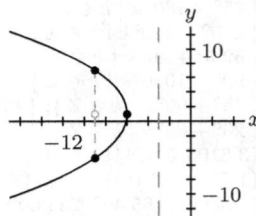

Set 137

1A. $x^2/(2)^2 + y^2/(3)^2 = 1$
B. $(x+1)^2/(1/2)^2 + (y-2)^2/(1/5)^2 = 1$
C. $(x-4)^2/(\sqrt{7})^2 + (y-1)^2/(\sqrt{3})^2 = 1$
D. $x^2/(1/5)^2 + y^2/(1/4)^2 = 1$
E. $(x-1)^2/(8)^2 + (y+5/2)^2/(2)^2 = 1$
F. $(x+3/2)^2/(121/20) + (y+5)^2/(121/8) = 1$

2A. $(x-1)^2/16 + (y-6)^2/25 = 1$
B. $(x+3)^2/289 + (y-5)^2/64 = 1$
C. $(x+10)^2/49 + (y+4)^2/25 = 1$
D. $(x-2)^2/9 + (y-6)^2/36 = 1$
E. $x^2/12 + y^2/3 = 1$ F. $x^2/2 + y^2/50 = 1$
G. $x^2/576 + y^2/676 = 1$
H. $(x+1)^2/100 + (y-4)^2/19 = 1$

3. $-2, 2$ 4. $-4\sqrt{3}/3, 4\sqrt{3}/3$

5A.   B.   C.

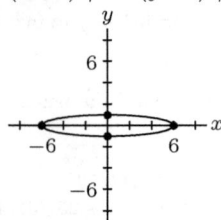 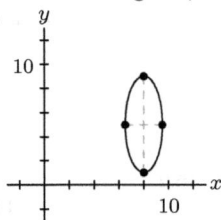 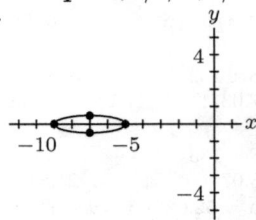

D.

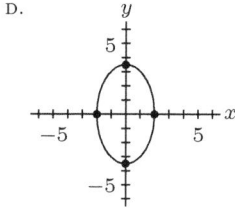

E.

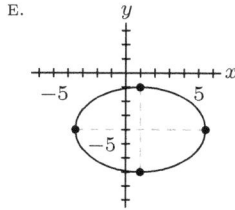

F.

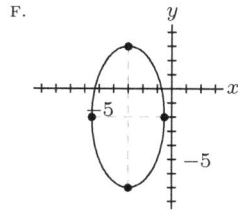

6A. foci: $(0, -3\sqrt{5})$, $(0, 3\sqrt{5})$;
vertices: $(0, -7)$, $(0, 7)$;
center: $(0, 0)$;
eccentricity: $3\sqrt{5}/7$;
length of major axis: 14;
length of minor axis: 4

B. foci: $(0, -2\sqrt{15})$, $(0, 2\sqrt{15})$;
vertices: $(0, -8)$, $(0, 8)$;
center: $(0, 0)$;
eccentricity: $\sqrt{15}/4$;
length of major axis: 16;
length of minor axis: 4

C. foci: $(-2, 7)$, $(6, 7)$;
vertices: $(-3, 7)$, $(7, 7)$;
center: $(2, 7)$;
eccentricity: $4/5$;
length of major axis: 10;
length of minor axis: 6

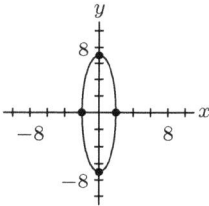

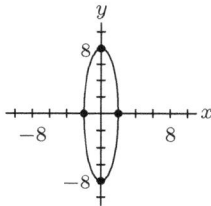

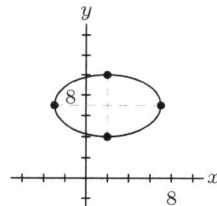

D. foci: $(-6, -2 \pm 2\sqrt{30})$
vertices: $(-6, -13)$, $(-6, 9)$;
center: $(-6, -2)$;
eccentricity: $2\sqrt{30}/11$;
length of major axis: 22;
length of minor axis: 2

E. foci: $(-3\sqrt{5}, 0)$, $(3\sqrt{5}, 0)$;
vertices: $(-9, 0)$, $(9, 0)$;
center: $(0, 0)$;
eccentricity: $\sqrt{5}/3$;
length of major axis: 18;
length of minor axis: 12

F. foci: $(-4, 0)$, $(4, 0)$;
vertices: $(-5, 0)$, $(5, 0)$;
center: $(0, 0)$;
eccentricity: $4/5$;
length of major axis: 10;
length of minor axis: 6

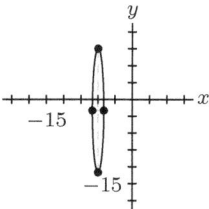

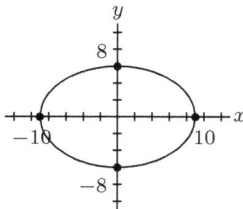

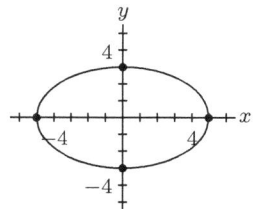

G. foci: $(-5 - 2\sqrt{21}, 0)$, $(-5 + 2\sqrt{21}, 0)$;
vertices: $(-15, 0)$, $(5, 0)$;
center: $(-5, 0)$;
eccentricity: $\sqrt{21}/5$;
length of major axis: 20;
length of minor axis: 8

H. foci: $(8, -3 - \sqrt{3})$, $(8, -3 + \sqrt{3})$;
vertices: $(8, -3 - \sqrt{5})$, $(8, -3 + \sqrt{5})$;
center: $(8, -3)$;
eccentricity: $\sqrt{15}/5$;
length of major axis: $2\sqrt{5}$;
length of minor axis: $2\sqrt{2}$

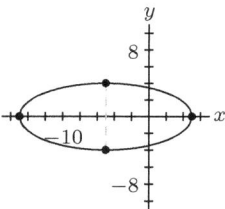

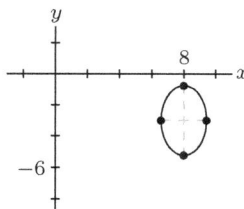

Set 138

1A. $(y-2)^2/(1)^2 - (x+9)^2/(1)^2 = 1$

 B. $(x-5/2)^2/(3)^2 - (y-1/2)^2/(3)^2 = 1$

 C. $(x+3)^2/(2\sqrt{3})^2 - (y-2)^2/(2)^2 = 1$

2A. $(x-2)^2/(3)^2 - (y-2)^2/(4)^2 = 1$

 B. $(y-9)^2/(3)^2 - (x-8)^2/(3\sqrt{3})^2 = 1$

 C. $(x+3)^2/(2)^2 - (y-6)^2/(5)^2 = 1$

 D. $(y-1)^2/(3)^2 - (x-8)^2/(4)^2 = 1$

 D. $(x+5)^2/(4)^2 - (y-3)^2/(4)^2 = 1$

 E. $(y-3/2)^2/(7/2)^2 - (x-4)^2/(7/2)^2 = 1$

3. $-15, 15$ 4. $-7\sqrt{3}, 7\sqrt{3}$

 E. $(y+4)^2/(7/2)^2 - (x+9)^2/(5/2)^2 = 1$

 F. $(x-0)^2/(\sqrt{3}/2)^2 - (y-11)^2/(5)^2 = 1$

 G. $(y+7)^2/(4)^2 - (x+8)^2/(20/3)^2 = 1$

 H. $(x+2)^2/(3)^2 - (y-5)^2/(12)^2 = 1$

5A.

B.

C.

D.

E.

F.

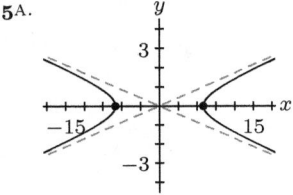

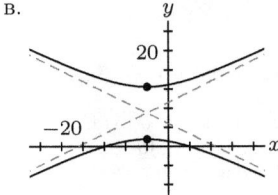

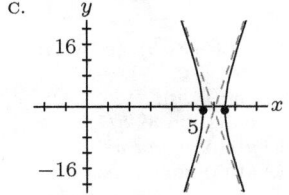

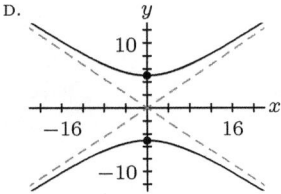

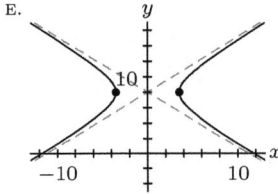

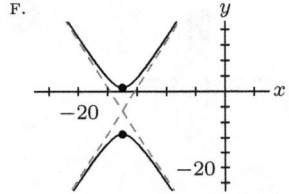

6A. foci: $(-3 \pm 3\sqrt{10}, 10)$;
vertices: $(-6, 10)$, $(0, 10)$;
center: $(-3, 10)$;
eccentricity: $\sqrt{10}$;
length of trans. axis: 6;
length of conj. axis: 18

B. foci: $(-6\sqrt{2}, 0)$, $(6\sqrt{2}, 0)$;
vertices: $(-8, 0)$, $(8, 0)$;
center: $(0, 0)$;
eccentricity: $3\sqrt{2}/4$;
length of trans. axis: 16;
length of conj. axis: $4\sqrt{2}$

C. foci: $(-5, \pm 3\sqrt{6}/5)$;
verts: $(-5, -\frac{2}{5})$, $(-5, \frac{2}{5})$;
center: $(-5, 0)$;
eccentricity: $3\sqrt{6}/2$;
length of trans. axis: 4/5;
length of conj. axis: $2\sqrt{2}$

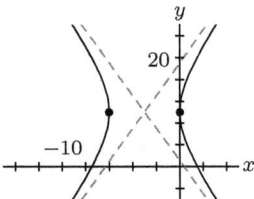

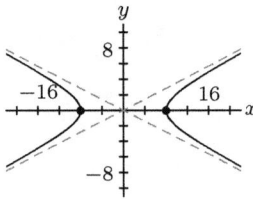

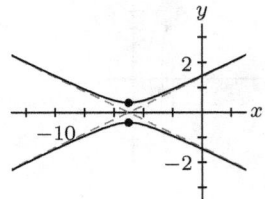

D. foci: $(6, 7 \pm \sqrt{165}/22)$;
vertices: $(6, 7 \pm \sqrt{11}/11)$;
center: $(6, 7)$;
eccentricity: $\sqrt{15}/2$;
len. of tr. axis: $2\sqrt{11}/11$;
length of conj. axis: 1

E. foci: $(11, -4 \pm \sqrt{7})$;
vertices: $(11, -4 \pm \sqrt{6})$;
center: $(11, -4)$;
eccentricity: $\sqrt{42}/6$;
len. of trans. axis: $2\sqrt{6}$;
length of conj. axis: 2

F. foci: $(-\sqrt{13}, 0)$, $(\sqrt{13}, 0)$;
vertices: $(-3, 0)$, $(3, 0)$;
center: $(0, 0)$;
eccentricity: $\sqrt{13}/3$;
length of trans. axis: 6;
length of conj. axis: 4

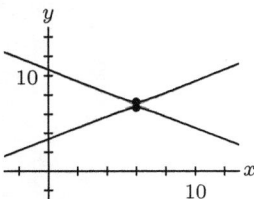

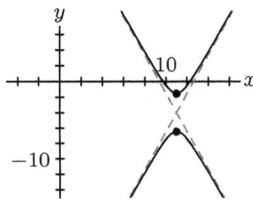

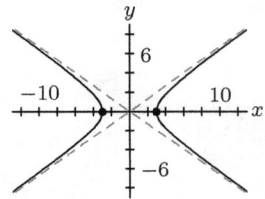

234

G. foci: $(2 - 2\sqrt{10}, 8)$, $(2 + 2\sqrt{10}, 8)$;
vertices: $(-3, 8)$, $(7, 8)$;
center: $(2, 8)$;
eccentricity: $2\sqrt{10}/5$;
length of transverse axis: 10;
length of conjugate axis: $2\sqrt{15}$

H. foci: $(0, -1 - 3\sqrt{5}/2)$, $(0, -1 + 3\sqrt{5}/2)$;
vertices: $(0, -5/2)$, $(0, 1/2)$;
center: $(0, -1)$;
eccentricity: $\sqrt{5}$;
length of transverse axis: 3;
length of conjugate axis: 6

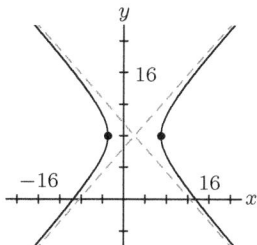

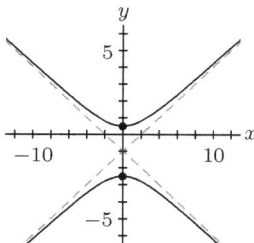

Set 139
1A. $(16, -40)$ B. $(-\sqrt{3}, -2\sqrt{3}/3)$ C. $(1, 1/2)$ D. $(2, 25)$
2A. $x = 3y - 12$ B. $x = (y + 5)^2 - 8$ C. $y = (x + 6)^3$

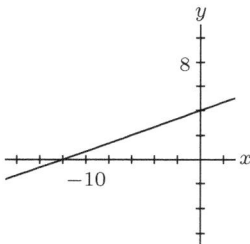

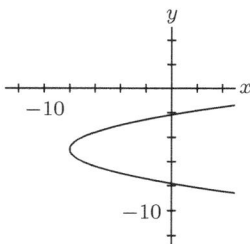

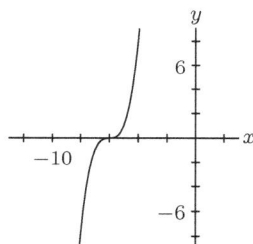

D. $x^2 + y^2 = 100$

E. $y = -|x - 9| + 1$

F. $y = -1/x$ for $x > 0$

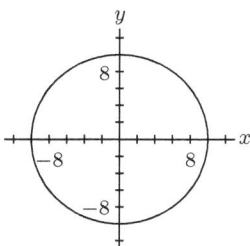

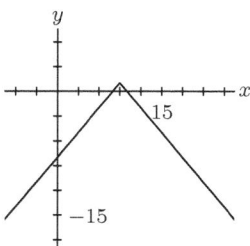

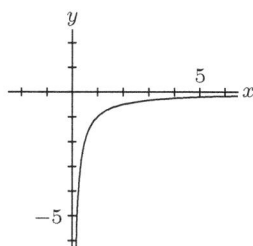

G. $\dfrac{(x - 2)^2}{(5)^2} + \dfrac{(y + 6)^2}{(11)^2} = 1$

H. $\dfrac{(x + 1)^2}{(6)^2} - \dfrac{(y - 4)^2}{(2)^2} = 1$

I. $y = 13 - x$

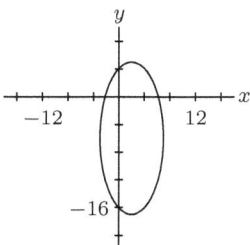

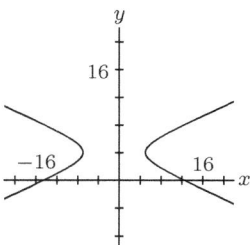

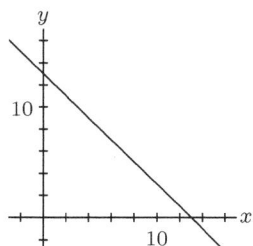

J. $y = (x-2)^2 - 4$

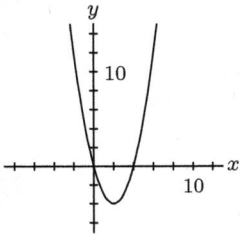

K. $x = |y - 3|$

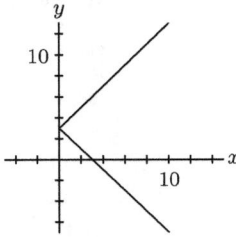

L. $y = 3 - x$

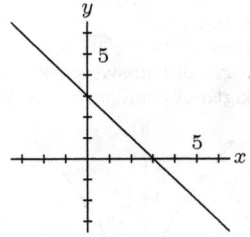

M. $x^2 + y^2/(3)^2 = 1$

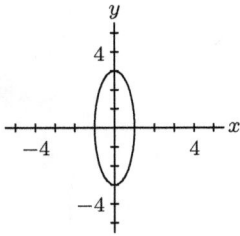

N. $y = \sqrt{x}$ for $x > 0$

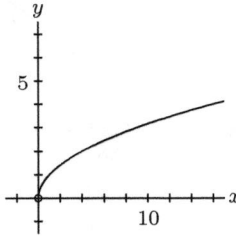

O. $(x-8)^2 + (y+4)^2 = 9$

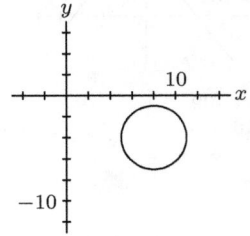

P. $\dfrac{(y-2)^2}{(4)^2} - \dfrac{(x-5)^2}{(9)^2} = 1$

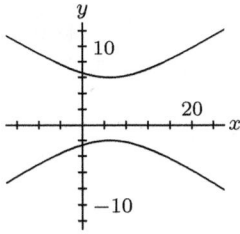

Q. $xy = 1$

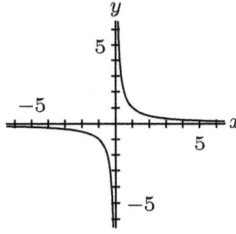

R. $xy = -1$

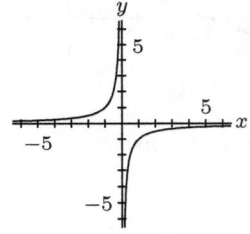

Set 140

1.

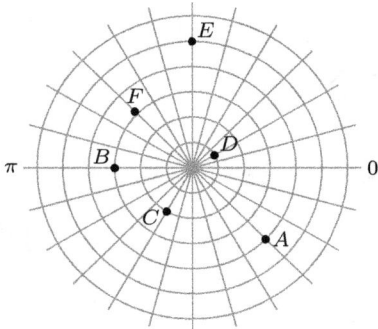

2A. $(8\sqrt{2}, 3\pi/4)$
B. $(4, 3\pi/2)$
C. $(4\sqrt{3}, 7\pi/6)$
D. $(5, 0)$
E. $(2, \pi/3)$
F. $(4, 5\pi/3)$

3A. $(5\sqrt{2}/2, 5\sqrt{2}/2)$
B. $(0, -7)$
C. $(-5, 5\sqrt{3})$
D. $(-\sqrt{3}, -1)$
E. $(-6, 0)$
F. $(0, 4)$

4A. $(6, 10\pi/3)$, $(-6, \pi/3)$
D. $(11, 3\pi/2)$, $(-11, 5\pi/2)$
C. $r = -3/(9\cos\theta + 4\sin\theta)$
F. $r = 10/(\cos\theta - 3\sin\theta)$

B. $(4, 7\pi/6)$, $(-4, \pi/6)$
5A. $r = 4$
D. $r^2 = -49/(\cos 2\theta)$
G. $r^2 = 10/(\sin 2\theta)$

C. $(1, 9\pi/4)$, $(-1, 5\pi/4)$
B. $r = 2/(\sin\theta)$
E. $r = 7/(\cos\theta)$
H. $r^2 = 25/(2 + \cos^2\theta)$

6A. $x^2 + y^2 = 64$; circle   B. $y = -x\sqrt{3}$; line   C. $(x-2)^2 + y^2 = 4$; circle
 D. $y = -7$; line   E. $x - y = 4\sqrt{2}$; line   F. $2x^2 + 7y^2 = 30$; ellipse
 G. $(x-6)^2 + (y+2)^2 = 40$; circle   H. $y = x$; line   I. $x^2 + y^2 = 25$; circle
 J. $x^2 + (y-7)^2 = 49$; circle   K. $x = 3$; line   L. $y = -x\sqrt{3} - 16$; line
 M. $y^2 - 2x^2 = 11$; hyperbola   N. $(x-3)^2 + (y-1)^2 = 10$; circle

Set 141

1A. 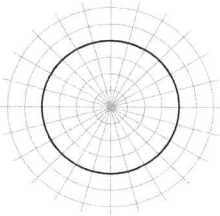   B. 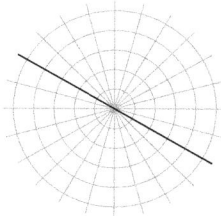   C.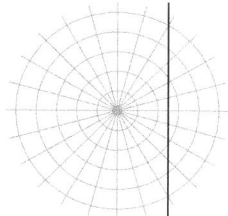

D. 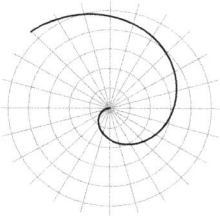   E. 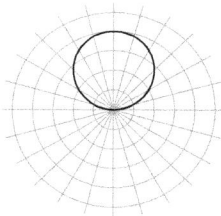   F.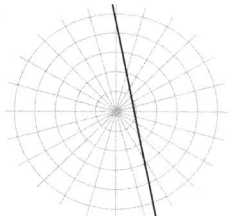

G. 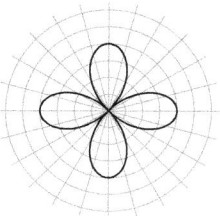   H. 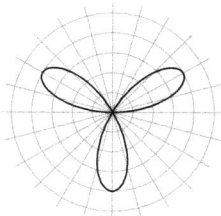   I.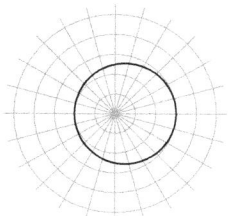

J. 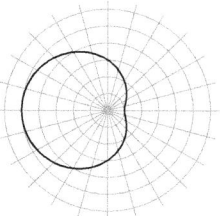   K. 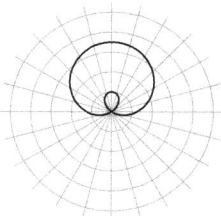   L.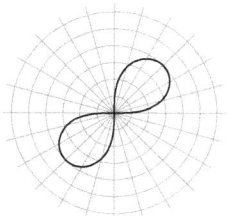

M. 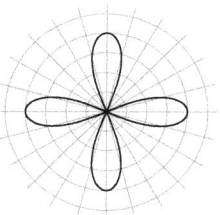   N. 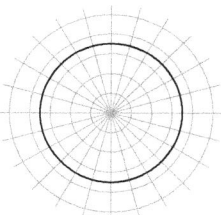   O.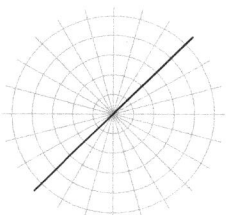

P.

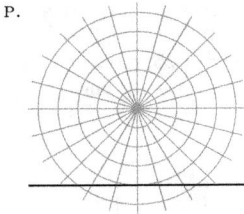

Q.

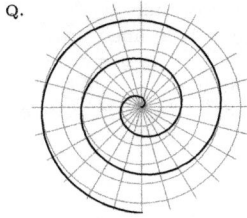

R.

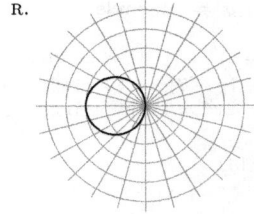

S.

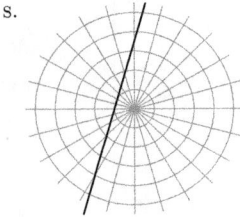

T.

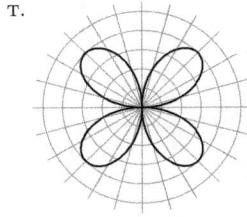

U.

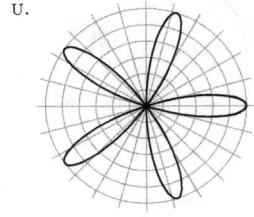

V.

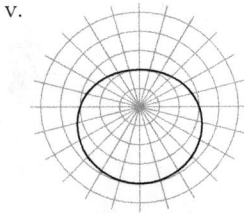

W.

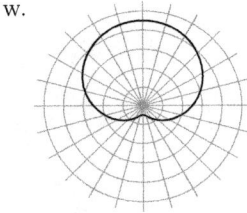

X.

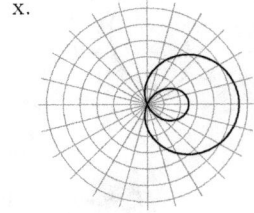

Y.

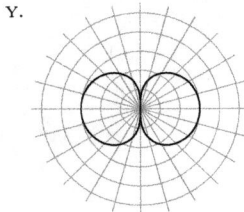

Z.

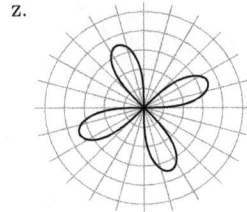

Set 142

1A. $(-\sqrt{2}/2, -11\sqrt{2}/2)$

2A. $(-5 + \sqrt{3}, -1 - 5\sqrt{3})$

B. $((1 + 8\sqrt{3})/2, (-8 + \sqrt{3})/2)$

B. $((-5\sqrt{6} + 2\sqrt{2})/2, (2\sqrt{6} + 5\sqrt{2})/2)$

3A. hyperbola;
$\cos 2\theta = 0$, $\cos \theta = \sqrt{2}/2$,
$\sin \theta = \sqrt{2}/2$, $\theta = 45°$;
$X^2/36 - Y^2/36 = 1$

B. parabola;
$\cos 2\theta = 0$, $\cos \theta = \sqrt{2}/2$,
$\sin \theta = \sqrt{2}/2$, $\theta = 45°$;
$Y = (\sqrt{2})X^2 - (3/\sqrt{2})/2$

C. ellipse; $\cos 2\theta = -24/25$,
$\cos \theta = \sqrt{2}/10$,
$\sin \theta = 7\sqrt{2}/10$, $\theta = 81.9°$;
$X^2/(236/9) + Y^2/4 = 1$

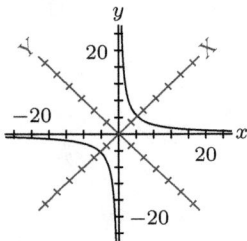

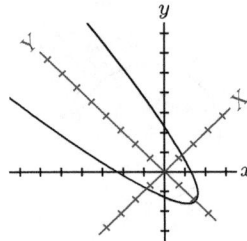

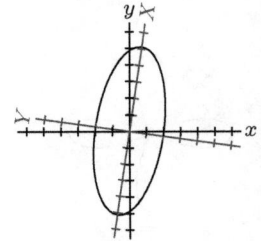

D. parabola;
$\cos 2\theta = -3/5,$
$\cos \theta = 1/\sqrt{5},$
$\sin \theta = 2/\sqrt{5},$
$\theta = 63.4°;$
$Y = -(X+5)^2 + 2$

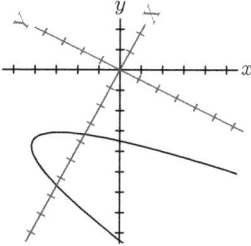

E. ellipse;
$\cos 2\theta = -15/17,$
$\cos \theta = 1/\sqrt{17},$
$\sin \theta = 4/\sqrt{17},$
$\theta = 76.0°;$
$(X+3)^2/8 + Y^2/4 = 1$

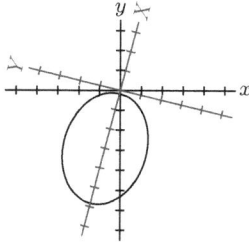

F. hyperbola;
$\cos 2\theta = 4/5,$
$\cos \theta = 3/\sqrt{10},$
$\sin \theta = 1/\sqrt{10},$
$\theta = 18.4°;$
$(Y-3)^2 - (X+5)^2/4 = 1$

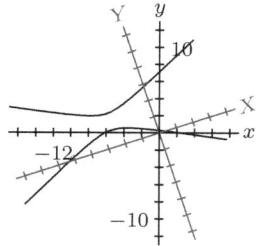

G. ellipse;
$\cos 2\theta = -1/2,$
$\cos \theta = 1/2,$
$\sin \theta = \sqrt{3}/2,$
$\theta = 60°;$
$\dfrac{(X-8)^2}{100} + \dfrac{(Y+1)^2}{20} = 1$

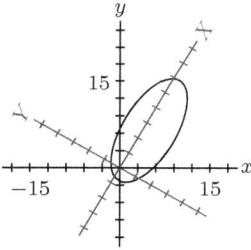

H. hyperbola;
$\cos 2\theta = 5/13,$
$\cos \theta = 3/\sqrt{13},$
$\sin \theta = 2/\sqrt{13},$
$\theta = 33.7°;$
$\dfrac{(X+1)^2}{4} - \dfrac{(Y+3)^2}{22} = 1$

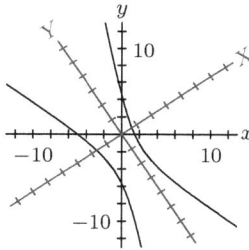

I. parabola;
$\cos 2\theta = -1/9,$
$\cos \theta = 2/3,$
$\sin \theta = \sqrt{5}/3,$
$\theta = 48.2°;$
$X = 2(Y-3)^2 - 7$

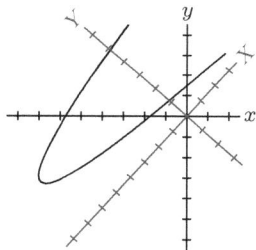

Set 143

1A. $-5 + 9i$; real: -5; imag.: 9
B. $12 + (-6)i$; real: 12; imag.: -6
C. $0 + (-4)i$; real: 0; imag.: -4
D. $3 + 0i$; real: 3; imag.: 0
E. $2 + i$; real: 2; imag.: 1
F. $1 + (-5)i$; real: 1; imag.: -5
G. $0 + 7i$; real: 0; imag.: 7
H. $-10 + 0i$; real: -10; imag.: 0

2A. $12i$ B. $(3\sqrt{2})i$ C. $(10\sqrt{2})i$ D. $9i$ E. $(2\sqrt{6})i$ F. $(6\sqrt{2})i$

3A. $7 + 19i$ B. $-15 - 19i$ C. $-5 + 31i$ D. $-18 - 99i$ E. $20 - 8i$ F. 65
G. $6 + 8i$ H. $13 - 6i$ I. $-33 + 54i$ J. $-48 + 20i$ K. $104 - 8i$ L. -41

4A. -6 B. $-\sqrt{10}$ C. $(12\sqrt{3})i$ D. $(2\sqrt{21})i$ E. -8 F. $-3\sqrt{5}$
G. $-(28\sqrt{3})i$ H. $(5\sqrt{14})i$ 5A. -1 B. $-i$ C. i D. 1

6A. $9 - 2i$ B. $4 + 3i$ C. $-1 - 7i$ D. $12i$ E. 8   F. 0
G. $-8 + 5i$ H. $-2 - 11i$ I. $10 + 6i$ J. $-4i$ K. -7 L. i

Set 144

1A. add.: $-2 - 4i$; mult.: $(1/10) - (1/5)i$
B. add.: $-8 + i$; mult.: $(8/65) + (1/65)i$
C. add.: 12; mult.: $-1/12$
D. add.: $-7i$; mult.: $-(1/7)i$
E. add.: $-5 + 3i$; mult.: $(5/34) + (3/34)i$
F. add.: $-9 - 2i$; mult.: $(9/85) - (2/85)i$
G. add.: -17; mult.: $1/17$
H. add.: i; mult.: i

2. $i, -i$ 3A. $m = 5, n = -8$ B. $m = -13/5, n = -9/2$

239

Set 145

1A. $5 - 2i$ B. $-15 + 9i$ C. $3 - 10i$ D. $-9 - 11i$ E. $7 + 4i$ F. $2 - 14i$
G. $-9 - 2i$ H. $-12 + 8i$ 2A. $(34/29) - (31/29)i$ B. $(-6/17) + (44/17)i$
C. $(-7/6) + (-13/6)i$ D. $(7/3) + 6i$ E. $(-1/25) + (32/25)i$
F. $(-7/41) + (-19/41)i$ G. $5 - 11i$ H. $(-3/2) + (-13/4)i$

Set 146

1A. $\{(3/2) - (\sqrt{7}/2)i, (3/2) + (\sqrt{7}/2)i\}$ E. $\{(-5/2) - (\sqrt{3}/2)i, (-5/2) + (\sqrt{3}/2)i\}$
B. $\{(-1/2) - (1/2)i, (-1/2) + (1/2)i\}$ F. $\{(7/6) - (\sqrt{11}/6)i, (7/6) + (\sqrt{11}/6)i\}$
C. $\{-3i, 3i\}$ G. $\{-4i, 4i\}$
D. $\{1 - (\sqrt{10}/5)i, 1 + (\sqrt{10}/5)i\}$ H. $\{(-3/4) - (\sqrt{3}/4)i, (-3/4) + (\sqrt{3}/4)i\}$
2A. -7; two distinct complex solutions   E. 0; one repeated real solution
B. 172; two distinct real solutions   F. -72; two distinct complex solutions
C. 0; one repeated real solution   G. 1; two distinct real solutions
D. 132; two distinct real solutions   H. -19; two distinct complex solutions

Set 147

1A. -8 (mult.: 2), -4 (mult.: 7), 0 (mult.: 5), 2 (mult. 6)
B. -2 (mult.: 4), 3 (mult.: 9), 4 (mult. 1), 10 (mult.: 6)
C. -11 (mult.: 5), 17 (mult.: 1), $-8i$ (mult. 3), $8i$ (mult.: 3)
D. 0 (mult.: 22), 1 (mult.: 2), 3 (mult.: 6), $-4 - 7i$ (mult. 5), $-4 + 7i$ (mult.: 5)
2A. $f(x) = -5x^3 + 20x^2 - 245x + 980$
B. $f(x) = x^7 - 4x^6 + 11x^5 - 32x^4 + 26x^3 + 20x^2 + 72x - 144$
C. $f(x) = 3x^4 + 12x^3 + 12x^2 + 48x$
D. $f(x) = -2x^6 + 32x^5 - 166x^4 + 236x^3 + 388x^2 - 768x - 720$
3A. $f(x) = 3(x + 4i)(x - 4i)$ B. $f(x) = (x + 1 + i\sqrt{3})(x + 1 - i\sqrt{3})$
C. $f(x) = 2(x - 3)(x + (3/2) + (3\sqrt{3}/2)i)(x + (3/2) - (3\sqrt{3}/2)i)$
D. $f(x) = (x - 3)(x + 5i)(x - 5i)$ E. $f(x) = (x + 5)(x - 2 + 3i)(x - 2 - 3i)$
F. $f(x) = -2(x + 3i)(x - 3i)$
G. $f(x) = 7(x - (5/14) + (\sqrt{3}/14)i)(x - (5/14) - (\sqrt{3}/14)i)$
H. $f(x) = -3(x + 2)(x - 1 + i\sqrt{3})(x - 1 - i\sqrt{3})$
I. $f(x) = (x + 7)(x + 2i)(x - 2i)$ J. $f(x) = (2x - 1)(x + 6 + i)(x + 6 - i)$
K. $f(x) = (3x + 2)(x + 1)(x - 2 + i\sqrt{14})(x - 2 - i\sqrt{14})$
L. $f(x) = (x - 1)(x - 4)(x + 3 + i)(x + 3 - i)$

Set 148

1.

Set 149

1A. $(6\sqrt{2}) + (6\sqrt{2})i$ B. $(5/2) - (5\sqrt{3}/2)i$ C. -7
D. $(-3\sqrt{3}/2) - (3/2)i$ E. $8i$ F. $(-3\sqrt{2}) + (3\sqrt{2})i$
2A. $4(\cos(3\pi/2) + i\sin(3\pi/2))$ B. $3(\cos(2\pi/3) + i\sin(2\pi/3))$ C. $10(\cos(5\pi/4) + i\sin(5\pi/4))$
D. $11(\cos 0 + i\sin 0)$ E. $14(\cos(\pi/6) + i\sin(\pi/6))$ F. $2(\cos(5\pi/3) + i\sin(5\pi/3))$

3A. $z_1 z_2 = 75(\cos(13\pi/12) + i\sin(13\pi/12))$;
$z_1/z_2 = 3(\cos(7\pi/12) + i\sin(7\pi/12))$

C. $z_1 z_2 = 6(\cos 45° + i\sin 45°)$;
$z_1/z_2 = 2(\cos 125° + i\sin 125°)$

B. $z_1 z_2 = 40(\cos(7\pi/6) + i\sin(7\pi/6))$;
$z_1/z_2 = (1/2)(\cos(3\pi/2) + i\sin(3\pi/2))$

D. $z_1 z_2 = 36(\cos 300° + i\sin 300°)$;
$z_1/z_2 = 9(\cos 80° + i\sin 80°)$

4. $z_1 z_2 = 12(\sqrt{6} + \sqrt{2}) + 12(\sqrt{6} - \sqrt{2})i$; $z_1/z_2 = (3(\sqrt{2} - \sqrt{6})/4) + (3(\sqrt{2} + \sqrt{6})/4)i$

Set 150

1A. $8(\cos(3\pi/4) + i\sin(3\pi/4))$
B. $625(\cos(4\pi/3) + i\sin(4\pi/3))$
C. $121(\cos(4\pi/3) + i\sin(4\pi/3))$
D. $-972 - 972i$

E. $81(\cos(2\pi/3) + i\sin(2\pi/3))$
F. $49(\cos(5\pi/3) + i\sin(5\pi/3))$
G. $81(\cos 0 + i\sin 0)$
H. -64

2A. $\{2i, \sqrt{3} - i, -\sqrt{3} - i\}$

B. $\{(5\sqrt{2}/2) + (5\sqrt{2}/2)i, (-5\sqrt{2}/2) + (5\sqrt{2}/2)i,$
$(-5\sqrt{2}/2) - (5\sqrt{2}/2)i, (5\sqrt{2}/2) - (5\sqrt{2}/2)i\}$

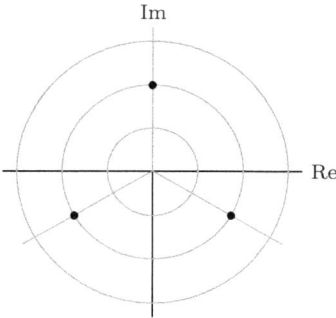

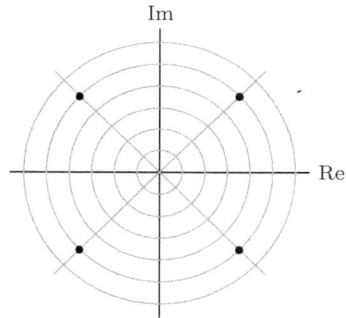

C. $\{\sqrt{3}, (\sqrt{3}/2) + (3/2)i, (-\sqrt{3}/2) + (3/2)i,$
$-\sqrt{3}, (-\sqrt{3}/2) - (3/2)i, (\sqrt{3}/2) - (3/2)i\}$

D. $\{\sqrt{3} + i, -1 + i\sqrt{3}, -\sqrt{3} - i, 1 - i\sqrt{3}\}$

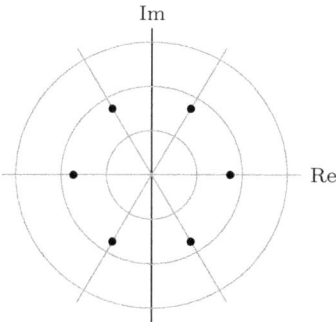

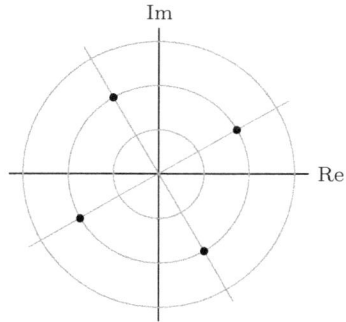

3A. $\{1, (\sqrt{2}/2) + (\sqrt{2}/2)i, i, (-\sqrt{2}/2) + (\sqrt{2}/2)i, -1, (-\sqrt{2}/2) - (\sqrt{2}/2)i, -i, (\sqrt{2}/2) - (\sqrt{2}/2)i\}$

B. $\{-\sqrt{3} + i, \sqrt{3} - i\}$

C. $\left\{ \dfrac{\sqrt{6}}{2} + \dfrac{\sqrt{6}}{2}i, \dfrac{\sqrt{6} - 3\sqrt{2}}{4} + \dfrac{\sqrt{6} + 3\sqrt{2}}{4}i, -\dfrac{\sqrt{6} + 3\sqrt{2}}{4} - \dfrac{\sqrt{6} - 3\sqrt{2}}{4}i, \right.$

$\left. -\dfrac{\sqrt{6}}{2} - \dfrac{\sqrt{6}}{2}i, -\dfrac{\sqrt{6} - 3\sqrt{2}}{4} - \dfrac{\sqrt{6} + 3\sqrt{2}}{4}i, \dfrac{\sqrt{6} + 3\sqrt{2}}{4} + \dfrac{\sqrt{6} - 3\sqrt{2}}{4}i \right\}$

D. $\{1 + i\sqrt{3}, -2, 1 - i\sqrt{3}\}$

E. $\{((\sqrt{6} - 3\sqrt{2})/4) + ((\sqrt{6} + 3\sqrt{2})/4)i, ((-\sqrt{6} + 3\sqrt{2})/4) - ((\sqrt{6} + 3\sqrt{2})/4)i\}$

F. $\{(3\sqrt{2}/2) + (3\sqrt{2}/2)i, (-3\sqrt{2}/2) + (3\sqrt{2}/2)i, (-3\sqrt{2}/2) - (3\sqrt{2}/2)i, (3\sqrt{2}/2) - (3\sqrt{2}/2)i\}$

4A. $\{\sqrt{2} + i\sqrt{2}, -\sqrt{2} - i\sqrt{2}\}$
B. $\{3 - 2i, -3 + 2i\}$

C. $\{3\sqrt{2}/2 - 3i\sqrt{2}/2, -3\sqrt{2}/2 + 3i\sqrt{2}/2\}$
D. $\{5 + 3i, -5 - 3i\}$

About the Author

The author was born in North Dakota and resides presently in Oregon. Other books by the author include:

Algebra and Trigonometry

Economics

A Study in Romans

Bimetrical Psalter